Insectpedia

Insectpedia

A Brief Compendium of Insect Lore

Eric R. Eaton

Illustrations by Amy Jean Porter

PRINCETON UNIVERSITY PRESS
Princeton & Oxford

Published by Princeton University Press
41 William Street, Princeton, New Jersey 08540
6 Oxford Street, Woodstock, Oxfordshire OX20 1TR

press.princeton.edu

ISBN 978-0-691-21034-6
ISBN (e-book) 978-0-691-23663-6

British Library Cataloging-in-Publication Data is available

Editorial: Robert Kirk and Abigail Johnson
Production Editorial: Mark Bellis
Text and Cover Design: Chris Ferrante
Production: Steve Sears
Publicity: Matthew Taylor and Caitlyn Robson
Copyeditor: Lucinda Treadwell

Cover, endpaper, and text illustrations by Amy Jean Porter

This book has been composed in Plantin, Futura, and Windsor

Printed on acid-free paper. ∞

Printed in China

10 9 8 7 6 5 4 3 2 1

Preface

This book is positively biased. Countless other books, plus magazine articles, advertising, videos, and social media memes demand we fear and exterminate insects. *Insectpedia* aims to refute those fears and ignite an appreciation of insects and those who study them. Even species for which we hold the most contempt, like tsetse flies, have redeeming qualities that will be illuminated here. Along the way, major principles of entomology will sneak in between the entertaining stories, and biographies of entomologists.

Insects are proof that evolution and instinct trump intelligence. We are overwhelmed by their sheer numbers, and forever a step behind in our efforts to subdue their impact on our health, agriculture, and wealth. They exploit our every weakness with maddening efficiency. Our relationship to insects has never been strictly adversarial, but it seldom serves the interests of business and industry to remember that. Do we turn some species into villains in order to turn a profit? That is less a conspiracy theory and more a shrewd business plan and marketing strategy.

Above all, we view insects as competitors for our resources. "Pest" is how we describe any species that we perceive as infringing on our property, person, or profits. Nature does not recognize the concept of ownership. Remarkably, natural ecosystems are seldom as chaotic as the artificial ones we create. There is no question that insects cause human misery and mortality directly, through the transmission of microbes that cause fatal diseases. Is it necessary, however, to eradicate the vectors? Billions are spent in campaigns to fog mosquitoes into oblivion, yet these are exercises in futility. In the United States, municipal spraying programs are driven by liability: the fear of lawsuits should citizens contract a disease and the city did nothing to prevent it.

Then there are "murder hornets," Spotted Lanternfly, and other pests that we have manufactured through accidental or intentional transport to foreign lands. This appears to be an acceptable price to pay for unfettered global commerce, our insatiable thirst for exotic landscape plants and cheap products, all packaged in containers that are themselves vulnerable to infiltration by pests. These illegal aliens are tolerated better than human refugees seeking asylum.

Climate change is also impacting insects. It is ironic that dire warnings of an "insect apocalypse" have finally generated recognition that insects provide essential ecosystem services that we, and all other living organisms, cannot survive without.

On a positive note, never has there been greater potential to turn the tide. The internet has made public access to entomologists easier than ever, and it has afforded scientists and nonscientists alike opportunities to contribute to our collective understanding of insects. Digital cameras and mobile phones allow us to capture photos, videos, and audio recordings. We can upload our observations to social media for the public to marvel at, and for entomologists to use in their research. Participation in citizen science projects further benefits various species and their habitats. We can become Master Gardeners and Master Naturalists to help others restore native plant communities, or at least modify our own yards and gardens to be wildlife-friendly. National Moth Week, and Fourth of July butterfly counts have made "bugwatching" a social endeavor. Butterfly houses and insect zoos bring exotic tropical insects to cities and towns near you.

It is impossible to have even a passing interest in the insect world and ever be bored. Something new and fascinating is revealed daily, either to you, personally, to the global community, or to both. From A to Z, the words I associate with insects are "amazing," "marvelous," and "zoophilia." Once you are finished with this book, dear reader, I hope you will be of a similar mind. If not, there are plenty of stories for which there was no room in this volume, and more discoveries await. Are you ready?

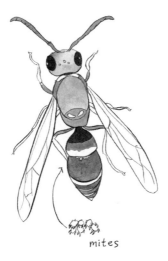

mites

Mason wasp
Ancistrocerus antilope

A carinaria

Some solitary bees, and mason wasps, have modifications to their anatomy designed exclusively for housing mites. An acarinarium is basically a "carport" where mites park themselves on the body of the insect. German entomologist Walter Karl Johann Roepke coined the term in 1920.

Many mites are parasitic, but the ones carried in acarinaria are mostly scavengers or fungivores. The wasps and bees are essentially transporting a cleaning crew to their nests, which are often located in linear tunnels divided into individual cells. Upon arrival at the nest, the mites disembark and commence feeding on materials that would pose a threat to the bee or wasp's egg, larva, or pupa offspring, or the pollen or prey cached for it. In the case of the wasp *Allodynerus delphinalis*, the mite *Ensliniella parasitica* functions as a bodyguard of immature *Allodynerus* by chasing off or assassinating the tiny parasitoid wasps *Melittobia acasta* before they can kill the *Allodynerus* larvae or pupae.

Acarinaria take various forms. Some large carpenter bees (*Xylocopa* spp.) have a concave chamber on the anterior (front) face of the abdomen where mites gather in a communal group. Mites on some mason wasps have more deluxe accommodations, nearly every mite having its own "garage" located at the front of the second tergite (dorsal, or top, abdominal segment) and covered by the rear edge of the first tergite.

Aerial Plankton

The next time you look out the window of an aircraft, know that you are not alone. The air outside is full of an astonishing diversity and abundance of insects and other arthropods. Serious study of atmospheric insect life took flight in the late 1920s, but even with airplanes available, London entomologist John L. Freeman was attaching traps to kites and collecting specimens in England and the United States. P. A. Glick created a device to attach to planes and made systematic surveys in Louisiana, and also Mexico, from 1926 to 1931, mostly 20–15,000 feet above the ground. Charles Lindbergh added a little data thanks to sticky glass slides affixed to the *Spirit of St. Louis* that he piloted across the Atlantic Ocean in 1933, flying mostly between 2,460 and 5,410 feet above sea level. In 1961, J. L. Gressitt placed a trap on a Super-Constellation aircraft that flew 116,684 miles specifically sampling the air. An alate (winged) termite, of all things, was collected at 19,000 feet.

Since then, little attention was paid to high-altitude bugs until the field of "aeroecology" was formally established in 2008. Locusts have been netted at 14,754 feet (4,500 meters). Aphids and related true bugs are surprisingly abundant, even at heights greater than 16,404 feet (5,000 meters). Small flies such as frit flies (family Chloropidae), pomace flies (Drosophilidae), and dark-winged fungus gnats (Sciaridae) have been recorded at more than 19,685 feet (6,000 meters). Thrips (order Thysanoptera), barklice (Psocodea), small parasitoid wasps (Hymenoptera), and small beetles (Coleoptera) also cavort in this aerial environment. The evolution of radar technology, especially Doppler "lidar," has assisted greatly in expanding our knowledge of both the quantity of arthropods present in the atmosphere, and how they move according to weather conditions.

"Albino" Insects

People unfamiliar with insect metamorphosis frequently describe any white insect they encounter as an "albino." Immediately after ecdysis (molting), many insects appear pure white as the new exoskeleton is soft, and pigments have not yet manifested. In others, especially true bugs, the freshly-molted insect may be pink or orange instead.

Some arthropods that live their entire lives in the deep recesses of caves, or in the soil, will also lack pigment, but that is not the same thing as albinism. True albinism is a genetic condition resulting in the complete lack of pigment in species where the normal default *is* pigmentation.

There are truly albino insects, but they are either rare or a product of laboratory breeding, or both. They are known in the migratory grasshopper *Locusta migratoria*, and in *Drosophila* pomace flies (though it is a yellow mutation similar to albinism). Complicating matters, there are some sulphur butterflies (family Pieridae) that can have white females instead of the normal yellow color.

Whiteflies, true bugs in the family Aleyrodidae, and many other insects for that matter, are covered in waxy white dust (pruinosity) or filaments, designed to make them unpalatable to predators, help prevent dehydration, reflect the heat of sunlight, or all of the above. Therefore, being a white insect does not equate to being an "albino" bug.

See also Exuviae; Metamorphosis.

Amber

Fossilized plant resins offer a literal window into the world of prehistoric insects. There are frequently multiple organisms within amber specimens, having been trapped in sap flows when the trees were living. We covet the gemlike quality of amber with or without entombed bugs, but entomologists are constantly revising the classification of insects in part because of specimens found in the durable, translucent substance. Other types of insect fossils are two-dimensional at best, but amber often presents a complete 360° view, or nearly so. That is, if gas bubbles or debris do not obscure crucial anatomical details.

This type of fossil is called an "inclusion" because the entire insect, not a facsimile of it, is preserved. Amber deposits are not uniformly distributed around the globe. The richest concentrations are in the Baltic region of Europe, in Myanmar, and in the Dominican Republic, though other sites occur in England, Austria, Lebanon, Jordan, and Japan. In most instances, the amber is derived from coniferous trees, especially pines. Dominican amber is from broad-leaved leguminous trees. Amber dates from the Holocene, our current geologic epoch, back to the Carboniferous period roughly 299–359 million years ago. It is remarkable how many ancient insects are easily recognized because we are familiar with their nearly identical present-day descendants.

By the way, amber is not always . . . well, amber. It comes in a variety of other colors such as gold, butterscotch, and rarely green or blue. Citizen scientist entomologists looking for a way to beat the winter blues might consider hunting for relict insects at the local jewelry store, or at gem shows, or online retail outlets. Who knows, you might make a valuable discovery.

See also Florissant Fossil Beds.

Anting

One of the most bizarre behaviors in the animal world is performed by birds which seek out the nests of ants for purposes that remain somewhat speculative. The fact that "anting" does not always involve ants only deepens the mystery.

More than two hundred species of birds display anting behavior, with twenty-four species of ants documented as "tools." Active anting involves the bird grasping one or more ants in its beak and applying them to feathers all over its body, especially the underside of the wings. In passive anting, the bird sprawls atop an ant nest, wings extended, and tail fanned, for maximum exposure to the aggravated insects. Since ants are a preferred food of few birds, there must be some other reason for birds to seek out ant colonies.

Many theories have been floated to explain anting, the most ludicrous of which suggests that it is for sexual gratification, given that some emphasis is given to the urogenital area of avian anatomy when employing ants in active anting. Another hypothesis is that the application of biting ants helps relieve pain and itching of new feather growth after molting.

The most likely explanation, backed up by anecdotal evidence and loose experimentation, is that the formic acid and other potent secretions from ants repel or kill mites, lice, and other ectoparasites. This idea also fits with anting behaviors that substitute beetles, true bugs, grasshoppers, earwigs, wasps, and millipedes for ants. Birds have also been seen grooming with rinds of citrus fruits, onion, beer, mustard, hair tonic, mothballs, even cigarette butts, and burning matches. The common denominator for all of these is one or more strong chemical components, if not insecticidal properties.

Aposematism

Brilliant "warning colors" abound in the insect world. They are usually advertisements of how toxic or well-defended the creature is, alerting would-be predators to keep their distance. They are so effective as a deterrent that perfectly harmless, tasty insects have adopted them as a ruse. The default fashion statement is usually a combination of black or iridescent blue with contrasting yellow, orange, red, or white.

Many insects will amplify the message by gathering together en masse. This congregating behavior is especially common in certain true bugs (order Hemiptera), caterpillars, sawfly larvae, and some wasps and bees. Nymphs of true bugs may molt synchronously to keep pace with the current wardrobe of their siblings, which may change with each instar (interval between molts).

Some insects combine aposematism with camouflage, relying on a cryptic appearance until the last possible moment, then flashing warning colors to startle the potential threat into retreating. The sudden display also gives the insect a brief moment to flee as the attacker contemplates its options. Mantids, stick insects, grasshoppers, and katydids are well known for such theatrics. They may even add an acoustical component to their act, like rattling their wings or scraping the abdomen against the wings to produce an ominous rhythmic sound.

Since aposematic colors are aimed mostly at diurnal, visual, vertebrate predators like birds and mammals, nocturnal insects need a different alert strategy. Tiger moths, while spectacularly colorful, emit warning clicks when flying at night to deter bats from eating them. Fireflies, glowworms, and railroad worms, which are beetles in the families Lampyridae and Phengodidae, use bioluminescence to advertise their toxicity. Indeed, even though not all adult fireflies flash or glow, the larva stage of every known species does so.

See also Bioluminescence; Stridulation.

archy the Cockroach

"There is always some little thing that is too big for us." Such is the wisdom of archy, a fictional cockroach created by writer, poet, and playwright Don Marquis. The roach, and his partner and foil, Mehitabel the cat, were introduced in Marquis's column in the New York *Evening Sun* in 1916. Supposedly, archy would race across the typewriter keys overnight leaving Marquis a manifest or verse to read in the morning. Since archy could not hit the shift key and a letter character at the same time, everything was in lower case, including his name. Punctuation was left in the dust, too.

The cockroach and cat enjoyed a popular and entertaining run through the early 1930s. It was an ingenious way for Marquis to write potent commentary on the issues of the day, as well as critique humanity as a whole, without risking much retaliation. After all, archy wrote it.

Archy the Cockroach

Mehitabel

after Don Marquis

So enduring are the characters of archy and Mehitabel, so delightful their humor, and so timeless their appraisal of our species, that they continue to be franchised in other media and one-person performances to this day. Various artists have set the prose and poetry to music, beginning in 1953, with actress Carol Channing as Mehitabel, and actor Eddie Bracken as archy. Bracken then teamed with Eartha Kitt, composer George Kleinsinger, and Mel Brooks to produce the Broadway musical *Shinbone Alley* in 1957, based largely on Marquis's work, with added dialogue by Brooks. Literary collections of the works can still be had, from *archy and Mehitabel* (1927) to *The Annotated Archy and Mehitabel* (2006). More to come?

See also Bugfolk.

Arthropods

The word "arthropod" suffers from popular mispronunciation as "anthropod." Perhaps this is because we hear more about anthropology, the study of humans and their culture and societies, than we hear about Arthropoda, the phylum of spineless wonders that includes insects. It could be argued that arthropods are far more significant than people, given that they account for the greatest biomass and biodiversity on planet Earth. Arthropods range in size from fairyfly wasps that could fly through the eye of a needle to gargantuan spider crabs with a leg span of 4 meters (12 feet) and weighing up to 45 pounds.

This extreme degree of diversity makes it difficult to generalize about arthropods, yet they share several characteristics in common. All arthropods have an external skeleton (exoskeleton). They lack a backbone, and so are invertebrate animals. Many species are so alien-looking that the public finds it hard to believe they are animals at all. Arthropods are usually bilaterally symmetrical, as we are, the left side mirroring the right. They do not, however, have a closed circulatory system. The blood, or hemolymph of an arthropod, bathes the entire body cavity.

The most significant feature of arthropods is the segmented body and jointed appendages. In fact, the word arthropod is modern Latin for "jointed foot." The phylum Arthropoda was erected by German zoologist Karl Theodor Ernst von Siebold in 1848, using the ancient Greek *árthron* for joint and *poús* for foot. Insects are arthropods in the class Insecta. Spiders, mites, ticks, scorpions, solifuges, harvestmen, horseshoe crabs (as of 2019), and related orders are in the class Arachnida. Millipedes are in the class Diplopoda, and centipedes in the class Chilopoda. Crabs, shrimp, lobsters, isopods, and other mostly marine arthropods are in the class Crustacea. Enjoy your next plate of seafood.

Autotomy

Some insects take self-sacrifice to an extreme. In the interest of their survival, most orthopteroids (stick insects, mantids, cockroaches, grasshoppers, katydids, and crickets) will happily break a leg if that body part is what a predator has a grip on. This self-amputation is called autotomy. Lizards that can sever their tails are another example. The loss of one limb barely slows an insect, and if it is not in the adult stage, it can grow a new one anyway the next time it molts its exoskeleton.

How does it work? The legs of insects that can willfully autotomize have a weakened fracture plane located at the trochanter-femur joint (think hip and thigh). Sustained stress beyond that point causes the leg to break free. Perhaps the most extreme form of autotomy has been documented in cave crickets in South Australia. The insect will eat its own hind leg in times of extreme food scarcity.

Honey bees, many social wasps, and some harvester ants (*Pogonomyrmex* spp.) have barbed stings that hold fast once embedded, and they tear out the venom sac and vital organs as the doomed insect departs.

A less drastic derivative of autotomy is autohemorrhaging, or "reflex bleeding," practiced by a number of common adult beetles, including ladybird beetles, leaf beetles, fireflies, and blister beetles. The hemolymph (blood) of blister beetles contains a potent defensive chemical called cantharidin. The hemolymph of other reflex-bleeders may be toxic or gooey and repel or entangle ants and other small predators. Autohemorrhaging is initiated, sustained, and terminated by changes in hydrostatic pressure.

See also Camel Crickets.

Beaded Lacewings

One of the more amusing anecdotes from the insect world is the case of the beaded lacewings, members of the family Berothidae in the order Neuroptera, which includes the familiar antlions, green lacewings, and brown lacewings. Adult neuropterans are lovely, mostly delicate insects, but their larvae are often voracious predators of other insects. Immature berothids are no exception, but those of *Lomamyia latipennis* have a peculiar and humorous means of dispatching their victims.

Hatching from an egg laid by its mother on a termite-infested log, the tiny larva searches feverishly for termites. Encountering one, it backs up to the termite and waves its rear end in the face of the much larger insect. In one to three minutes, the termite keels over, completely paralyzed. It has apparently fallen victim to the fatal flatulence of the *Lomamyia* larva, which commences feeding on it. The beaded lacewing larva spends two or three weeks behaving in this manner before molting into a sedentary second instar (interval between molts). This couch potato stage lasts only a couple of days before another molt turns it back into a termite-killing machine.

The larger larva has the ability to conquer as many as six termites at once with its ass-first attack style. The "fart" consists of an allomone that apparently poisons termites only, as other insects in the vicinity, like barklice, are unaffected even in an enclosed chamber in a laboratory. So far, *Lomamyia latipennis* is the only species known to use this predatory technique. *Lomamyia hamata* simply skewers a termite with its mandibles and injects a paralytic neurotoxin, like most other lacewing larvae. Why termites do not recognize beaded lacewing larvae as a threat remains an unsolved mystery.

Beer Bottle Beetles
Habitat loss and climate change are among the leading causes of insect population declines, but in the case of the western Australian jewel beetle *Julodimorpha*

bakewelli, littering almost did them in. Thankfully, this story has a reasonably happy ending. The male beetles were disproportionately drawn to discarded glass "stubbies" of a certain brand of beer that had the same texture and patterns of reflected light as a female beetle, and were similarly colored. This kind of attraction is termed a "supernormal stimulus," as the inanimate bottle was much larger than the female beetle, but bore all the right love signals. The real female is indeed larger than the male, and flightless, so both beetle and bottle were encountered in similar circumstances in the arid environment where the beetles breed.

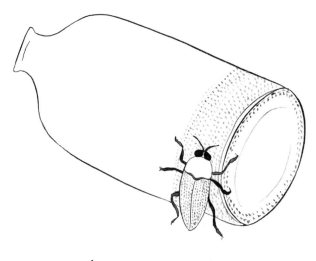

Australian jewel beetle
Julodimorpha bakewelli

This was no laughing matter for the beetles, who put forth so much futile energy in attempted copulation with the bottles that they died of exhaustion, dehydration, or were killed by ants. The phenomenon was first documented in a photo in 1980, prompting field experiments by Darryl Gwynne and David Rentz the following year. The two were later awarded an infamous "Ig Nobel Prize" in biology in 2011 for their efforts. A more respectful nod came from breweries who changed the design of their bottles to eliminate the textured area that was attracting the beetles. The beetle itself did make it onto a one-dollar Australian postage stamp released on September 6, 2016, one of four jewel beetles so honored.

Biblical Bugs

One literary source that continues to be responsible for our attitudes toward insects is the *Holy Bible*. Though there are plenty of verses that recognize insects as examples of industriousness and humility, the overwhelming interpretation of most passages is that insects are intolerable enemies bringing pestilence and hunger. Lice, flies, and locusts were among the ten plagues in Exodus, for example.

Historically, these plagues have been interpreted as punishments sent by God for human immorality and sin. In scientific light, they can be viewed simply as recurring natural phenomena, or the result of poor attention to sanitation, and other earthly causes, no divine intervention involved. Locust swarms continue to be a periodic issue, leaving starvation and poverty in their wake. Climate change and increasing human populations conspire, perhaps, to create more frequent locust plagues with more intense effects. The natural cycles responsible for El Niño warming events alone could trigger increases in populations of malaria-spreading mosquitoes and other disease-carrying flies.

The Bible is, overall, an anthropocentric document focused on human shortcomings and moral potential, with references to arthropods used accordingly. God admonishes us to "Go to the ant thou sluggard; consider her ways, and be wise." (Proverbs 6:6). The idea that instinctive behaviors can reflect moral character is also an intriguing notion. Meanwhile, British biologist and geneticist J.B.S. Haldane, in 1951, asserted that the Creator has "an inordinate fondness for beetles," based on Haldane's observations of biodiversity.

Biocontrol

The Agricultural Revolution brought unprecedented challenges to controlling pests. To this day crop losses can be substantial, owing to the scale of modern farming itself, and to the constant threat of foreign species introduced accidentally or intentionally. Biocontrol is the employment of natural enemies to control pests. This includes enlisting insect predators and parasitoids, herbivorous insects in the case of invasive plants, microbial and viral pathogens, and manipulating or interrupting hormonal cycles during insect growth.

In theory, biocontrol targets a specific pest without harming nontarget organisms. The practice dates back to at least the thirteenth century, when nests of weaver ants (*Oecophylla smaragdina*) were placed in citrus groves in China to reduce populations of a certain stink bug.

There have been spectacular success stories since then, such as the case of the Australian *Novius cardinalis* lady beetle saving California citrus groves from an outbreak of the Cottony Cushion Scale in the late 1800s.

The Industrial Revolution compounded the Agricultural Revolution, increasing scale by several orders of magnitude, and requiring new ways of addressing pests. The answer was broadcast spraying of chemical insecticides, especially after World War II, when advances in chemistry and production made them cost-effective solutions. Consequently, biocontrol languished in the United States until Rachel Carson's *Silent Spring* called into question the long-term effects of chemical treatments.

Great strides continue to be made in refining techniques for evaluating potential biocontrol organisms. Even so, many species introduced to control invasive insects and plants have had deleterious impacts on native species. The future of biocontrol remains optimistic, but will likely have to be conducted in tandem with cultural controls like crop rotation, and a return to smaller scale enterprises in a "retrolution" that allows experimentation with lower risk of financial crises.

See also Bt; Integrated Pest Management.

Bioluminescence

Glow-in-the-dark organisms have always captured our imagination, perhaps none more so than fireflies, beetles in the family Lampyridae. They are not the only insects to produce light, however, and, interestingly, it is mostly larval insects that light up, rarely the adults. How is such "cold" light produced? The overly simplified explanation is a chemical reaction in which the compound luciferin combines with oxygen as catalyzed by the enzyme luciferase. This phenomenon is different from fluorescence, where an organism glows under exposure to ultraviolet light.

There appear to be three purposes for bioluminescence in insects. One is a nocturnal version of aposematism, whereby bioluminescence serves to warn potential predators that the organism is toxic or distasteful. Many firefly species are loaded with lucibufagins, steroids akin to toad toxins. All known larval fireflies are luminescent. The other purpose for self-sparkling is to attract prey. Larvae of many species of predatory fungus gnats, family Keroplatidae, glow continuously after hatching from the egg, and spin tubular shelters of mucous-like silk. In *Arachnocampa luminosa* of New Zealand, a long strand of sticky silk issues from the larval retreat. The communal glow of hundreds of larvae draws small insects into the silken traps. The larva then ingests both the silk and the entangled victim. Several caves are major tourist attractions because of this unique insect.

Larvae of the click beetle *Pyrearinus termitilluminans* live in termite mounds in the Pantanal region of central Brazil. They glow at will from segments immediately behind their heads. When the termites liberate swarms of new, winged queens and males (alates), the glowing beetle larvae are an irresistible attraction. The immature beetles feast during these swarming events. Swarms of alate ants meet the same fate.

The third use of bioluminescence, to attract mates, is today viewed as a secondary adaptation.

See also Aposematism; Fluorescence.

Boll Weevil

Few insects are as despised and, ironically, celebrated as the Boll Weevil, *Anthonomus grandis*. The beetle single . . . snoutedly(?) brought the U.S. cotton industry to its knees in the early 1900s. Presumably native to central Mexico, it migrated here in the late 1890s. It has been spreading through South America, too, entering Brazil in roughly 1983.

The female beetle uses the jaws on the end of her beaklike rostrum to gnaw a hole in a cotton square (flower bud), or young boll. She then lays her eggs inside. The grubs that hatch commence feeding, destroying the square in the process. There can be several generations annually.

We have thrown nearly every control strategy in our arsenal at the beetle and still it persists. Initial attempts to thwart its spread by creating cotton-free zones met with so much political resistance that they were never implemented. Pesticides succeeded only in creating resistant weevils, and suppressing it enough to allow aphids and bollworms to rise to the rank of most wanted cotton pest.

Such a fixture is the Boll Weevil in the economic and cultural reality of the southern United States that a traditional blues song was written about it, and a monument erected. The song has nearly infinite variations and has been recorded by multiple artists since the 1920s. Brook Benton's version earned him three weeks at number 2 on *Billboard* magazine's "Hot 100" beginning in mid-July of 1961. The monument is placed in the town of Enterprise, Coffee County, Alabama. Dedicated on December 11, 1919, the original, built in

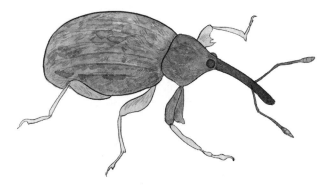

Boll weevil
Anthonomus grandis

Italy, featured a woman holding a shallow bowl. In 1949 the crowning adornment of the weevil on a pedestal was added, along with a fountain at the base. The entire piece is more than 4 meters (13 feet) tall, and a replica now stands in for the original to prevent permanent damage from vandalism or theft.

Bombardier Beetles

Many insects defend themselves chemically from predators, but bombardier beetles, a subset of ground beetles in the family Carabidae, deploy their deterrents explosively. Our understanding of the mechanism of chemical warfare in bombardiers is the result of decades of relentless curiosity, innovation, and collaboration driven by the late Dr. Thomas Eisner. Working with chemists, engineers, and Harold Edgerton, the pioneer of high-speed photography, Eisner pieced together the remarkable weaponry of this insect.

Adult *Brachinus* beetles, and related genera, have paired glands and reservoirs inside the abdomen. The glands manufacture chemicals called hydroquinones, which are stored together with hydrogen peroxide in the large reservoirs. When the beetle is molested by an adversary, muscles wrapping the reservoirs contract, forcing the liquid through a valve and into a reaction chamber immediately in front of the anus. The chamber contains two kinds of enzymes. One enzyme, a catalase, fractures the hydrogen peroxide into oxygen and water. The other, a peroxidase, oxidizes the hydroquinones into benzoquinones. The result is an explosion that blasts benzoquinone at the attacker.

Not only is the chemical irritating, it is hot, exiting the beetle's rear at 100° Centigrade. The spray can be delivered in any direction thanks to the articulation of the abdomen. Further, that one audible "pop" represents many pulsed micro-explosions. The bombardier possesses not a cannon, but a machine gun. Bombardier beetles are so remarkable that one was featured, at Eisner's urging, in a set of twenty commemorative U.S. postage stamps issued on October 1, 1999. Flip stones or logs, especially near water, and you may encounter one or more bombardiers. Think twice before grabbing one, though.

Braun, Annette Frances (1884–1978)

Pioneering female scientists are too often ignored instead of celebrated the way their male counterparts are.

Entomologist Annette Braun achieved many firsts in her lifetime, including being the first woman to receive a doctoral degree from the University of Cincinnati, Ohio in 1911. Her younger sister, Emma "Lucy" Braun was the second. The two siblings made an extraordinary team, Lucy studying plants while Annette excelled in the study of microlepidoptera, moths so small as to be overlooked by most entomologists today.

While they traveled mostly in Ohio, and later through Kentucky, the duo made thirteen trips to the western United States beginning in the 1930s. Annette possessed a critical eye for detail and a talent for scientific illustration. She published her first monograph in 1914, her last in 1972. Dr. Braun authored the scientific names of more than 340 new species in the course of her lifetime. In 1926, she was elected vice president of the Entomological Society of America; she was elected as a correspondent to the Academy of Natural Sciences of Philadelphia in 1949, and her collection of 30,000 leaf-mining moths, most of them reared at the sisters' home in Cincinnati, now resides there. Annette was also voted president of the Ohio Entomological Society in 1959.

Beyond their scientific endeavors and accomplishments, the Braun sisters were vocal conservationists, especially when it came to advocating protection for the unique, unglaciated habitats of Adams County, Ohio. Their legacy is reflected in preserves of remnant prairies and other ecosystems, and the state-of-the-art Eulett Center housing offices of The Nature Conservancy, and field laboratory of the Cincinnati Museum of Natural History.

Brochosomes

Have you ever noticed that some sharpshooter leafhoppers in the family Cicadellidae have white, chalky
patches on their front wings? No? Now you will have
to be on the lookout. It turns out the oval, convex spots
are not species-specific markings, but something more
remarkable.

Brochosomes are protein-lipid particles produced by
glands associated with the Malpighian tubules, part of
the excretory system of insects. Each particle usually
takes the form of a hollow sphere with a surface re-
sembling a honeycomb of hexagonal and pentagonal
"compartments." The overwhelming property of these
particles, especially when layered, is their extreme resistance to wetting. This could be advantageous when
crowding results in copious amounts of liquid waste
(honeydew) from individuals raining down on each
other. Accumulations of the sugary compounds can
mire the insects, sticking their body parts together.
Honeydew can quickly acquire mold spores that could
compromise the health of the insects, too. Even a dewy
morning in a meadow could trap a leafhopper, to say
nothing of an afternoon downpour.

Brochosomes appear to be products unique to ci-
cadellid leafhoppers, and widespread in their application by the insect to its external cuticle. The bug uses
its long hind legs to coat its body as we would do with
sunscreen. Females of the tribe Proconiini also produce
cigar-shaped brochosomes that they store as those ob-
long white badges, just prior to laying their eggs. Leafhopper females insert their eggs into plant tissues with
bladelike ovipositors. They then cover the wound with

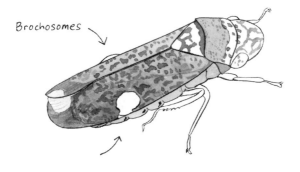

Brochosomes

Broad-headed sharpshooter
Oncometopia orbona

these brochosomes, perhaps to repel predators, prevent their eggs from molding, or both. Investigation into the role of brochosomes is in its relative infancy.

See also Honeydew; Malpighian tubules.

Brood Parasitism

Burying beetles, carrion beetles in the genus *Nicrophorus*, are best known for disposing of small mammal, bird, and reptile carcasses through burial, raising small numbers of offspring on the "meatball" they make out of the decedent. *Nicrophorus pusulatus* has broadened its *modus operandi* in a surprising way.

A long-term study of the Black Rat Snake, *Pantherophis obsoletus*, at the Queen's University Biological Station in Ontario, Canada, revealed something wholly unexpected. Many eggs excavated from nests were damaged or destroyed by *N. pustulatus*. Beetle larvae were found in some eggs, while small holes were chewed in others, probably by adult beetles to feed themselves or provide entry for their larvae. This was a startling case of brood parasitism of a vertebrate host by an invertebrate. Technically, because the beetle larvae always kill their host, they are classified as parasitoids, a term usually reserved for solitary wasps, and flies, that exploit other insects as hosts.

The snakes use communal nests in the same location, year after year, which makes them a dependable resource for the beetles. It allows the beetle to achieve greater reproductive success than other burying beetles. This appears to be a habitual strategy for *N. pustulatus* throughout its geographic range. Beetle larvae have also been observed in the nest of a Fox Snake in southern Ontario, and in rat snake nests in Illinois. How the species came to exploit this reptilian resource, and how the adult beetles locate snake eggs, remains unclear. Perhaps it all started with one beetle that happened upon broken or unhatched eggs.

The discovery of this phenomenon illustrates the need for more field biologists, and greater cooperation between scientists of different disciplines. A greater understanding of ecological relationships should be the driving force behind most field research.

Bt

Bacillus thuringiensis, abbreviated as "Bt," is one of the most heavily used biocontrol agents in pest management. Renowned for its ability to target specific pests, it has been a weapon of choice since 1970. Ironically, Bt was first regarded as an enemy of silkworms. In 1901, Japanese biologist Shigetane Ishiwatari isolated a bacterium he named *Bacillus sotto*, determined to be the cause of sotto disease in the caterpillars. Ten years later, Ernst Berliner rediscovered it in the Mediterranean Flour Moth, a stored product pest. Berliner named it *Bacillus thuringiensis* after the town of Thuringia, Germany where the moth occurred. That name stuck.

The bacterium kills with toxins produced by crystals in the microbe's spore stage. Bt requires ingestion of the spores by the insect in order to kill it. The typical chain of events begins when the insect's midgut dissolves the crystal, liberating proteins known as delta-endotoxins. Enzymes in the midgut activate the toxins, resulting in cessation of feeding by the insect, and erosion of its gut lining. Bt bacteria, and the normally helpful gut fauna of the insect, then spread throughout its body, leading to death.

Besides being applied like a conventional pesticide, Bt is incorporated into transgenic plants, whereby the gene responsible for programming the crystal protein toxins is introduced to the genome of the plant. The plant thus manufactures its own Bt toxins. Since 1996, cotton, soybean, corn, potato, and other crops with Bt genes have been grown around the world.

While Bt is as "organic" as one could hope for, it is imperfect. Applications must be timed precisely to coincide with the life cycle of the target pest. The right strain must be used, of which thousands exist. It also does not persist in the environment the way chemical treatments do. Resistance has been documented in a number of species, particularly the Diamondback Moth and Indian Meal Moth.

See also Biocontrol; Integrated Pest Management.

Bugfolk

Science detests anthropomorphism, the assignment of human emotions and purpose to other animals. Fortunately, for our entertainment, artists have no problem turning other beasts, including insects, into caricatures of people. The late entomologist Charles L. Hogue coined the term "bugfolk" in 1979 to describe these artistic mergers of man and insect.

Cartoonists often portray insects as bipedal, with four limbs instead of six, faces with rounder features, eyes with pupils, and clothed in human attire. Whether this makes insects more appealing, or people less so, depends on the artist's intentions. Many an editorial cartoonist has "ento-morphed" humans to embody our less-than-desirable attributes, or reflect the socio-political climate of the day. This is especially evident in the work of French artist J. J. Granville in the 1840s. British illustrator L. M. Budgen followed in Granville's footsteps, but was true to her own style, and cleverly published under the pseudonym *Acheta domestica*. Her three-volume *Episodes of Insect Life* alternated between the scientifically accurate and utterly whimsical.

The smoking caterpillar of Lewis Carroll's *Alice's Adventures in Wonderland* was rendered by John Tenniel in 1897. Jiminy Cricket, a Disney creation in 1940, exudes joy and positivity. The desire to pair human intellect with the physical prowess of insects is satisfied by superheroes like the Green Hornet and Ant-Man. Other artists have turned bugfolk into demons, fairies, and other supernatural entities.

Contemporary cartoonist Gary Larson has achieved fame in part by framing insect behavior with human

after L. M. Budgen, "Sipping their cups of dew"
Episodes of Insect Life, vol. 2, 1850

concerns of embarrassment and disappointment, and turning the idea of human superiority on its head. Taking insects out of normal context and placing them in corporate workplaces, suburban households, playgrounds, highways, and other human habitats magnifies the bugfolk persona. Larson's style and subject matter have inspired countless other cartoonists who took the spotlight after he retired. Recently, he has made a comeback via the internet.

See also archy the Cockroach.

Bushman Arrow Poison Beetles

Many insects are toxic to predators, but human exploitation of those species is rare. One example is the San people of the northern Kalahari Desert in southern Africa, who poison their hunting arrows with toxins from chrysomelid leaf beetles in the genera *Diamphidia* and *Polyclada*, and from the beetles' parasites, three species of the carabid beetle genus *Lebistina*.

The two *Diamphidia* species feed on foliage of *Commiphora* trees (myrrhs), while *Polyclada flexuosa* consumes leaves of *Sclerocarya birrea* (marula). Presumably, the larvae sequester compounds from these plants to manufacture their potent toxins. After reaching maturity, the larvae drop to the ground and burrow deeply (0.5–1 m). It is there that they may encounter the *Lebistina* larvae. The parasitoid attaches to the leaf beetle larva before it constructs the dense, spherical cell of soil in which it will pupate. Inside, a leaf beetle pre-pupa can live in diapause for two to four years.

Bushmen locate the host plants of these beetles and dig beneath them to find the pupal chambers. Breaking them open, they extract the pre-pupal larvae, or the pupae. The simplest method is to squeeze the insects so they secrete body juices over the arrow shaft near the arrowhead. Alternatively, plant juices, such as those of the toxic *Sansevieria aethiopica*, and human saliva are mixed and rubbed over the arrow; or the insects are dried, ground into a powder, and mixed with plant extract before application. Drying the finished product helps bond the poison to the arrow. The toxins can retain their lethal potency for at least one year. Interestingly, bushmen covet *Lebistina* larvae as most toxic of all.

Struck with a poisoned arrow, a large animal may succumb in a few hours, or take several days to die. The action of the toxin is foremost hemolytic, bursting blood cells and compromising the supply of oxygen to all cells in the body.

Camel Crickets

More spider-like than insect-like, members of the family Rhaphidophoridae tend to give people the chills. Some call them spider crickets or "sprickets" owing to their gangly appearance, but "camel cricket" is the name most often applied. The humped appearance of their compact, wingless bodies makes them easy to recognize. Females bear a sword- or knifelike ovipositor at the rear of the abdomen. Both sexes sport a pair of flexible appendages called cerci, also at the rear end. The fact that these insects haunt places like basements, cellars, wells, mine shafts, and caves only serves to enhance the cringe factor. Cave cricket is in fact another name for them. They are not, however, true crickets.

Some cave-dwelling species are endemic to only a single subterranean labyrinth. Most inhabit the twilight zone rather than the deeper recesses, but their extraordinarily long antennae help them navigate in the absence of light. In stark contrast, sand treaders have short, stubby legs, the hind ones equipped with "sand baskets" to help them dig rapidly into shifting dunes. In many urban areas, the common species is often the immigrant Greenhouse Camel Cricket, *Diestrammena asynamora*, native to Asia but now widespread thanks to international commerce.

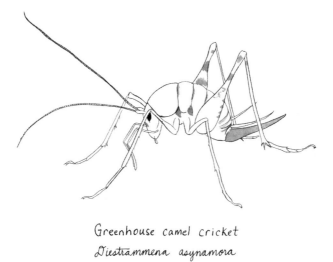

Greenhouse camel cricket
Diestrammena asynamora

Camel crickets are nocturnal and omnivorous, active at night and eating almost anything organic. A handful of species are occasional pests in mushroom-growing operations, but in nature they are critical members of the detritivore community that help recycle decaying matter into soil nutrients. Many of the forest- and cavern-inhabiting species appear to require some sort of social contact, huddling in groups of three or more during the day in crevices, animal burrows, or other shelter.

Chrysalis

The pupa stage of butterflies, and sometimes moths, is called a chrysalis, a Latin word derived from the Greek word *khrusos*, for "gold." The term is often confused with "cocoon," but the two are not interchangeable. A cocoon is a nonliving material that sometimes envelops the pupa stage of an insect. A caterpillar may wander far from its food plant and seek another substrate on which to pupate.

Being the "resting stage" of the butterfly life cycle, the chrysalis is typically an inert but living object highly vulnerable to predators and parasitoids. Consequently, a chrysalis can take any number of forms to conceal itself through camouflage. The typical swallowtail butterfly pupa may be disguised as a broken twig, for

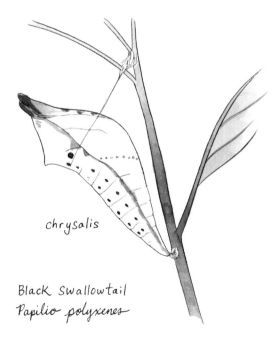

chrysalis

Black swallowtail
Papilio polyxenes

example. Chrysalides (plural of chrysalis) of some long-wing butterflies in the tribe Heliconiini of the family Nymphalidae resemble bits of withered foliage. The metallic markings on some chrysalids are more difficult to explain but may mimic reflections of sunlight on water droplets.

Inside, a chrysalis is anything but inactive. Cells are broken down from the larval body and resurrected as adult tissues. Genes are turned off, others turned on. This explains why the caterpillar of a Monarch butterfly doesn't attempt to *walk* to Mexico. The process is regulated chiefly by an array of hormones, and the absence of the juvenile hormone that preserved the youth of the

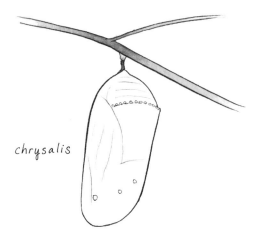

chrysalis

Monarch
Danaus plexippus

caterpillar stage. Everything must proceed flawlessly if a perfect adult butterfly is to finally burst through the chrysalis exoskeleton to freedom.

The chrysalis is an iconic symbol of metamorphosis and promise in human culture. It represents bottled-up potential, the precursor to something great and beautiful. It can also symbolize death and resurrection, as it did among ancient Egyptians.

See also Cocoon; Juvenile Hormone; Metamorphosis.

Cochineal

Some insects are to dye for. *Dactylopius coccus*, popularly known as cochineal, is even farmed for the sole purpose of supplying natural dye. That white fluff on prickly-pear cacti is wax secreted by cochineal scale insects. There are eleven species, collectively ranging from South America to the southwestern United States. The insects turn their diet of cactus sap, in part, into carminic acid. It serves the insect as a bitter defensive chemical that repels predators. It also gives the insect a vivid red color.

The insect itself was not identified until 1725, but it was valued as a dye as early as the second century BC by Aztecs and Mayans. When Spanish conquistador Montezuma enslaved the residents of several cities in the fifteenth century, the defeated metropolises were each ordered to pay an annual "tribute" of 2,000 dyed cotton blankets, and 40 bags of raw dye, the dried and powdered insects themselves. Cochineal became Mexico's second largest export during the colonial period. By the seventeenth century, the dye was traded widely throughout Europe and India.

The end of the Mexican Revolution in 1821 consigned an end to the nation's monopoly on cochineal, as farming the insects began in Guatemala, the Canary Islands, and Spain and Algeria in the 1830s. Production of synthetic dyes in the mid-nineteenth century crashed the cochineal industry, but when some commercial dyes were revealed to be carcinogenic, there was renewed interest in cochineal. Currently, Peru is the leading exporter of cochineal, followed by Chile and Mexico.

The principal use of cochineal today is as a laboratory stain in microbiology. It continues to be a traditional dye in Mexican folk crafts and is an ingredient in some cosmetics, and in coloring of food and pharmaceuticals.

Cocoon

A cocoon and a pupa are not the same thing. A cocoon is a covering that surrounds the pupa, usually made of silk spun from glands in the head of the larval insect prior to its entering the pupa stage. The larva may spend days, weeks, months or, in rare instances, years in diapause as a pre-pupa, before finally molting into the pupa stage. Besides moth and (a few) butterfly caterpillars, larvae of ants, bees, wasps, fleas, caddisflies, and lacewings, antlions, and their kin also spin cocoons.

The shroud of a cocoon offers the vulnerable insect protection in a number of ways. It insulates the organism from the elements. It makes a barrier to water and deters predators and parasitoids from gaining entry. Caterpillars decorated with irritating hairs or venomous spines may impregnate their cocoons with those setae for added protection. Some larvae incorporate living or

cocoon
Live oak tussock moth
Orgyia detrita

dead foliage into the cocoon for camouflage purposes. Still others cement soil particles with silk or saliva to make a hard, sturdy capsule. Larvae of social wasps that build paper combs simply cap their individual cell with a silken dome.

Emergence from a cocoon by the adult insect is no small feat. The pupa may be anchored to the cocoon at the rear to facilitate purchase as the soft adult strains to escape. Secretions of fluid serve to soften or dissolve the silk ahead of the insect.

Cocoons have always intrigued and inspired us. "Cocooning" suggests a state of comfort and safety, a way

cocoon

Giant silk moth
Hyalophera cecropia

to relieve stress. We have cocoon-like comforters, blankets, and sleeping bags, ironically made mostly from synthetic or plant-based fibers. Silkworms spin pristine cocoons of silk that is so valuable it led to a human enterprise of staggering significance.

See also Chrysalis; Gypsy Moth; Sericulture.

Collins, Margaret James Strickland (1922–1996)

The story of Margaret James Strickland Collins demonstrates exceptional perseverance in the face of racism and gender bias that still afflicts the field of entomology.

Dr. Collins was a gifted child, graduating high school at 14, then earning a Bachelor of Science in Biology at West Virginia State University. She enrolled in graduate classes at the University of Chicago and was mentored by Alfred Emerson, a world authority on termites, who still held her back from fieldwork. Collins persisted, and she received her PhD in Zoology, making her the first black female entomologist in the United States.

She joined Howard University as an assistant professor, but sex discrimination affecting timely promotions led her to seek full professorship at Florida A&M. There, her aspirations collided with racism. When invited to lecture at a neighboring college, a bomb threat directed at her forced cancellation of the engagement. Later, a bus boycott in Tallahassee, initiated by the Florida A&M Student Council, saw Dr. Collins volunteering as a driver for commuters.

Collins returned to Howard as a full professor in 1964 and also became a senior research associate at the Smithsonian. In a final irony, she collaborated with two students in Guyana to reopen the Alfred Emerson Research Station in 1979, named after her graduate school mentor.

Together with David Nickle, Dr. Collins discovered and described a new species of dampwood termite, *Neotermes luykxi*, published in 1989. Collins' collected insect specimens now reside at the National Museum of Natural History in a special collection bearing her name.

She also led a symposium on human equality at the American Association for the Advancement of Science in 1979. The proceedings were published in 1981 as *Science and the Question of Human Equality*, by Westview Press.

Comstock, Anna Botsford (1854–1930)

Behind every great man is a great woman, so the saying goes. Anna Botsford Comstock can stand on her own, thank you. Yes, she did marry her college professor, John Henry Comstock, and illustrate his works. She excelled at both wood engraving and pen and ink. Mrs. Comstock illustrated and cowrote the *Manual for the Study of Insects*, the standard text of entomology from 1895 through its twentieth edition in 1931. She was also an author, teacher, and conservationist. To his credit, John Henry never failed to champion his wife's accomplishments.

The classic *Handbook of Nature Study*, published in 1911, has gone through twenty-five editions, been translated into eight languages, and is still in print. It reflects Comstock's dedication to environmental literacy, and her belief that fundamental knowledge of ecology and environmental science were of paramount importance for proper stewardship of the Earth. The book was a high achievement in the nature study movement, but she also published curricula and traveled the country to instruct schoolteachers in the art of nature education.

Comstock reached high levels of education herself, being appointed to assistant professor of natural science at Cornell University in 1898, the first female in the teaching faculty. Some members of the board of trustees objected, however, and she was demoted to lecturer. Reinstated later, she was promoted to full professorship in 1919, just prior to her retirement. The university eventually immortalized both Anna and John by naming a building after them.

So profound was her impact that the National Wildlife Federation placed her in its Conservation Hall of Fame, proclaiming Anna Botsford Comstock the "Mother of Nature Education" in that 1988 ceremony. She was also the subject of the Entomological Society of America's Founders' Memorial Award Lecture in 2017, delivered by Dr. Carol M. Anelli, the eleventh female honoree. Comstock herself was the third female recipient.

Cricket Fighting

Staged battles between vertebrates are considered barbaric, but animal rights activists have yet to object to contests between invertebrates. Perhaps this is because most tournaments take place outside of North America; and arthropods of the same species and sex rarely kill or maim each other. Cricket fighting, with male *Gryllus* species, is restricted to China, where it enjoys a history dating back to the Tang Dynasty (618–907).

Cricket fighting is not trivial entertainment. A good deal of money can be made in betting on the outcomes of matches. Male crickets are bred for aggressiveness, the champion fighters capable of earning up to $20,000 over their adult life span and accruing national acclaim. Prime Minister Jia Sidao was so obsessed with cricket fighting that he wrote a how-to book for aspiring cricket owners. Some scholars believe his passion interfered with his ability to govern, leading to decline of the South Song Dynasty (1127–1279). Ming Xuan-Zhong (ca. 1427–1464) was known as the "Cricket Emperor" for his affinity for the sport. The communist government that held sway during the cultural revolution of 1966–1976 banned cricket fighting as an example of the leisure pastime of the elite class. The youth of today have embraced the bouts as part of a lost heritage.

Prizefighters in the cricket world can be created without a pedigree. While larger individuals tend to dominate, reaching their peak about twelve days into adulthood, other factors can enhance performance. Research shows that males that have just mated are more aggressive, as are individuals housed in isolation for a period of time. Males that have secured a fortress like a burrow or crevice are likewise antagonistic to other males. These nuances are not lost on owners of fighting crickets, and great care and ceremony is given to the sport, with investments in housing and feeding.

Darning Needles

Dragonflies are steeped in myth and superstition, which has led to many regional names for them based on such folklore. Among these epithets is "darning needle." Indeed, the common name for the family Aeshnidae is "darners," perhaps owing to the long, slender, somewhat tapering shape of the abdomen of these elegant aerialists.

The name carries a sinister reputation for the poor insect, especially when "devil" is attached, as in "devil's darning needle." According to folklore, dragonflies will sew up various body parts, such as lips, ears, nostrils, or eyelids. This fate mostly awaited naughty or untruthful children, but the New England version claims anyone falling asleep within reach of a dragonfly might have their fingers or toes stitched together.

As if that isn't frightening enough, other myths have dragonflies associating with serpents, referring to them as "snake doctor," "snake feeder," or "snake servant." Dragonflies supposedly alerted snakes to approaching threats, or helped them find food. Folks were thus discouraged from causing harm to dragonflies, lest a snake seek revenge.

Other aliases for dragonflies include "devil's riding horse," "horse-stinger," and "bullstang." From European cultural history comes "adder's needle," "adder's servant," "adderspear," and "ox-viper." Dragonflies cannot sting, but anyone who has seen a female darner ovipositing in a log, aquatic vegetation, or mud can understand how the act could be misinterpreted, and used to full advantage to scold impressionable youngsters into behaving themselves.

Leave it to Eastern culture to redeem the evil ascriptions of Western culture. Dragonflies are celebrated in Japanese culture as the embodiment of strength, courage, and happiness. They are also given spiritual significance as representatives of spirits visiting homes.

Deathwatch Beetles
The original head-banger might be the deathwatch beetle, *Xestobium rufovillosum*, of the family Ptinidae. The insect lives in tunnels it bores in wood, and adult

males and females find each other by slamming their faces on the floors of their tunnels, then orienting to the direction of the replies. This is seismic communication, as they do not hear the sounds, but perceive the vibrations instead. It is audible to us, though, prompting dire superstitions that the tick-ticking of the beetles was an omen of death in old houses infested by the insects.

The male beetle initiates dialogue by tapping four to eleven times in rapid succession. Females apparently tap only in response to males. Males may travel long distances, only to find they overshot their female target, or made some other mistake in orienting to her. Not all females are receptive to mating at all times, either, so may not respond. It is the beetle equivalent of Tinder: it works often enough to keep the population going.

In nature, deathwatch beetles bore into dead or dying trees, and they may infest lumber at any stage between logging and milling, or even later. Once established,

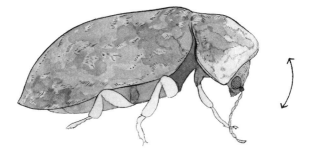

Deathwatch beetle
Xestobium rufovillosum

beetle populations can sustain themselves for many generations. They prefer old wood, and wood infested with fungi appears to be most coveted.

Old structures are thus prone to prolonged infestation by deathwatch beetles. In the absence of any human noise, the sound of lovelorn males can be all the louder. The eerie tapping gave rise to the belief that someone in the household would soon die. This might not be far-fetched if the timbers are so structurally undermined by generations of beetles that the building is in imminent danger of collapse.

See also Tok-Tokkies.

Delusory Parasitosis

The belief that "bugs" are infesting one's body, burrowing under the skin and causing unbearable irritation, is surprisingly common. It carries a stigma that the victim is to blame, and that they willfully draw an irrational conclusion. Delusional parasitosis is the clinical term for this malady, shortened to "delusory parasitosis."

Typically, four attributes indicate this condition. First, the person presents "evidence" of lint, dust, and other detritus, usually in a plastic baggie, box, or stuck to clear adhesive tape. The recipient of this material, be they entomologist or physician, is told that there are bugs in the sample. Invariably, there are not. The second indicator is an exhaustive narrative describing the offending creatures, and speculation as to their origin. The third warning sign is refusal to accept any alternative explanation for their symptoms. Lastly, the person may have inflicted damage to their skin in an effort to rid themselves of the sensations and "bugs."

There could be a host of other causes for their skin irritation. It could be connected to their behavior, prior circumstances, other illnesses, or drugs. Delusory parasitosis commonly afflicts methamphetamine addicts, for example. Someone who has had a previous infestation of cockroaches, lice, bed bugs, or other pests can be more prone to delusory parasitosis. Other medical conditions, and side effects of some prescription drugs, can produce similar symptoms.

In the absence of an obvious arthropod-related cause, the entomologist is left with no choice but to recommend evaluation by a physician and/or psychologist to eliminate other potential causes. Providing empathy while retaining professional protocol is not easy, but that should be the goal.

Diapause

In insects, "hibernation" is a misnomer. The correct term is diapause, which can be seasonal like vertebrate hibernation, but not necessarily so. Diapause is a temporary cessation of growth and development, and suppression of metabolic processes, which may be provoked by any number of conditions, including food scarcity.

In most insects, diapause is facultative, meaning it is sparked by environmental cues that initiate changes in bodily chemistry. Obligatory diapause is known for a few insects, in which case it occurs regardless of environmental stimulus, and *must* occur in order for growth and development to proceed normally. Most facultative diapause is triggered by changes in daylength, the usual precursor to extreme cold or heat. A period of chilling is usually required before an insect can exit from diapause, in regions with distinct seasons.

Many moths undergo diapause as caterpillars. Diapause of insect larvae in general is controlled by hormones, or their absence in the case of steroids that regulate molting. Larvae may resume feeding once their diapause is over; in other cases, they have been in diapause as a prepupa, and pupation is the next activity. The solution to food scarcity in some dermestid beetles, like *Trogoderma glabrum*, is not to enter diapause, but to molt regressively into a smaller larva, repeatedly if need be, and resume a normal course when favorable conditions return.

Adult diapause is expressed as a halt in reproduction, and sexual organs do not produce eggs or sperm during that interval. Migratory populations of the Monarch butterfly enter diapause upon reaching their wintering grounds. They subsist on fat reserves accumulated during the journey. Other insects, especially true bugs, beetles, wasps, and flies, also enter diapause as adults. Their wing muscles may shrink so there is less metabolic demand on their fat reserves.

See also Chrysalis; Cocoon; Juvenile Hormone; Metamorphosis.

Doodlebugs

Before the internet, and the distraction of cell phones, children had to work to find amusement, and it usually took them outdoors. One favorite pastime seems to have been pestering the larvae of antlions, members of the family Myrmeleontidae, and related to lacewings.

Look in dusty or sandy soil at the base of trees, under bridges, rock overhangs, and other spots sheltered from rain and you will likely find several small, funnel-like pits. These are produced by "doodlebugs," the whimsical name we have assigned to these deadly predators of ants and other insects. Tickle the walls or floor of the pit and you may get a rise out of the occupant buried at the bottom. Traditionally, a child would chant a verse like "Doodle-bug, doodle-bug, are you at home?" while prodding the pit. From the Caribbean to Asia, Australia, and South Africa, variations on this theme abound.

"Doodlebug" stems from the meandering scrolls left by an antlion larva seeking a new site for its pit. The plump, stubby insect walks backwards in a circle, spiraling inwards, in order to construct a pit. It flings sand beyond the perimeter, using its head as a shovel. Not all antlions dig pits. Most species bury themselves in the sand and wait in ambush with open jaws.

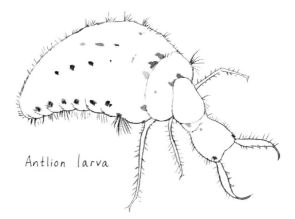

Antlion larva

Antlions pupate in silken cocoons at the bottom of their pits. They more than make up for the pudgy appearance of their youth, emerging as slender adults with two pairs of long, delicate wings. Watch for them at your porch light at night, or fluttering feebly through tall grass in fields. Meanwhile, check out *The Antlion Pit* website for online enchantment.

Drosophila "Fruit Flies"

If there is one insect responsible for the most scientific advances, it could well be the pomace flies in the genus *Drosophila* in the family Drosophilidae. Often referred to as "vinegar flies" or "fruit flies," the latter name correctly applies to members of the family Tephritidae.

These are the "gnats" (again a misnomer) that flit through your kitchen and alight on overripe bananas, the wine glass, and anything else with an odor of fermentation. The flies' attraction to, and tolerance for, all things alcoholic stems from the need to find appropriate food resources for their maggot offspring. The larvae thrive best on the yeasts that colonize decaying organic matter. The life cycle is very short, completed in ten days at 25° Celsius, and can quickly yield high populations of adult flies.

It is this rapid turnover of generations that has made *Drosophila* one of the most popular laboratory organisms, especially for genetic research. That, and the supersized chromosomes found in their salivary glands. It was pomace flies that aided Dr. Thomas Hunt Morgan in his research on heredity and helped win him the Nobel Prize in Physiology or Medicine in 1933. Since then, of the more than 2,000 known *Drosophila* species, *D. melanogaster* has continued to be the go-to lab fly. Publication of its genome sequence took place in the year 2000.

The flies will no doubt continue yielding new revelations in the fields of cell biology, developmental biology, neurobiology and behavior, population genetics, and evolutionary biology. What does that mean for you? Well, we *Homo sapiens* do share 60% of our DNA with *Drosophila melanogaster*, and 75% of the genes known to cause human diseases can be found in fruit flies.

Ecosystem Services

Humans have a tendency to justify everything in purely economic terms. What cannot be appraised in dollars and cents is all too often not valued at all. It is easy to calculate what insect pests cost us in damages, but the accounting of the benefits of insects is infinitely more difficult.

John Losey and Mace Vaughan, in a 2006 article in *BioScience*, selected pollination by native insects, pest control by predators and parasitoids, disposal of fecal material and decomposing plants and animal carcasses, and food for fish and wildlife as four major ecosystem services provided by insects. They estimated the economic value of these to be at least $57 billion annually in the United States alone. This estimate does not include pollination of crops by non-native honey bees, honey and other products from insects, nor seed dispersal by ants, maintenance of forest health through pruning of trees, and other tasks performed by insects that do not directly impact humanity. The authors also excluded species introduced and/or reared for the exclusive purpose of pest control in agricultural ecosystems.

This landmark paper finally elevated insects to their rightful place as economic engines, not economic disasters. It also served as a call for mitigation and restoration of natural habitats lost to agriculture and urbanization. Promoting biodiversity even on the fringes of plots and subdivisions increases the potential impact of beneficial insects in those adjacent human ecosystems. We could all use less expenditure for chemical pest control, and more fruits and vegetables resulting from pollination.

More recently, scholars have expanded beyond these support and regulatory services to include recognition of insects in provisioning services (food, silk, dyes, shellac, and other material resources), medicine, and cultural services such as recreation, tourism, and spiritual values.

See also Cochineal; Entomophagy; Lac Insect; Medicinal Maggots; Pollinators; Seed Dispersal; Sericulture.

Endangered Insects

The poster animal for endangered species conservation is never going to be an insect, but that does not mean that there are no bugs on the brink of extinction. The overwhelming threats to insects are the same as they are for vertebrate wildlife: habitat destruction and climate change. Pesticide use is an ongoing problem. Competition from invasive foreign species is of increasing concern.

The most critically imperiled insect species are those that occupy limited, unique habitats such as caves, bogs and other wetlands, isolated sand dune systems, ever-shrinking prairies, steadily warming alpine areas, and islands. There are many species endemic to such locations, found nowhere else on the planet, and by adaptation unable to exist anywhere else. Protection and preservation of those localized ecosystems is of paramount importance in securing the safety of the insects and other organisms that live there. Fail at this and all other efforts will be futile.

Captive breeding programs have been a strategy for saving many vertebrates, with the goal of returning a portion of the offspring to the wild. This strategy is being applied to a few insects, too. In the United States, several butterflies, including the Karner Blue, and beetles, including the American Burying Beetle and Salt Creek Tiger Beetle, are reared for release. Internationally, zoos create species survival plans for threatened animals of all kinds.

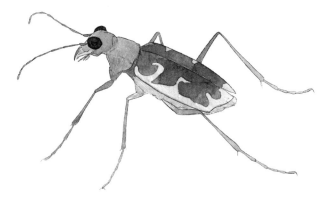

Salt Creek tiger beetle
Cicindela nevadica lincolniana

While extinction is a natural phenomenon, the *rate* of species extinctions is increasing almost exponentially as a result of human activities. Less than 1% of all known insect species have even been evaluated for potential listing as endangered or threatened at the national or international level.

See also Insect Apocalypse; Weta; Xerces Society.

Entomology

The study of insects is frequently confused with etymology, the study of words. Born largely out of innate curiosity, entomology has expanded and evolved to meet one crisis after another.

The Greek philosopher Aristotle began giving order and meaning to the physical world, but it was not until the Renaissance that René Antoine Ferchault de Réaumur of France (1683–1757) began studying the anatomy, function, metamorphosis, and other biological aspects of insects.

The 1800s saw an age of exploration and colonialism, with all specimens retained by colonizers. To this day we have scarcely extended reparations for the hoarding of collections. The travels of Darwin, Alfred Wallace, and others did, however, inspire important principles such as natural selection. Concurrently, the Industrial Revolution caused a dramatic increase in the scale of agriculture. Entomology turned its attention to pest control, and by the early 1900s some biological controls were being utilized to combat pests.

World War II forced entomologists to confront emergencies like typhus, a louse-borne disease. Triumph was achieved through development of insecticides like

DDT. After the war, chemical controls were applied to agricultural and urban pests. Victory was short-lived. Insects developed resistance to many insecticides, while vertebrates suffered. These shortcomings were brought into public view by Rachel Carson in *Silent Spring*, and by entomologist Robert van den Bosch in *The Pesticide Conspiracy*. The revelations led in part to creation of the U.S. Environmental Protection Agency in 1970.

Students pursuing entomology today have a wealth of career choices. Forensic entomology has re-emerged as a vital component of criminal science. Medical and veterinary entomologists address issues of public health. Forest entomologists manage native and invasive pests, as well as broaden our understanding of insect ecosystem services. Insect conservation is expanding to meet the challenges of decreasing insect abundance and diversity.

Entomophagy

The practice of eating insects has a long history that is experiencing something of a resurgence in our modern age. In Western cultures, the "ick factor" now struggles to reconcile with recognition of the nutritional value of insects.

There is little question that insects are an excellent source of protein, fat, and key minerals and vitamins. Many insects are exploited as food in parts of Africa, Asia, Australia, and Latin America. Consumption of insects has its roots in the traditions of indigenous peoples but has become a commercial enterprise in some places. The giant water bug *Lethocerus indicus* is so coveted that it commands high prices; this is especially

Mopane worms (fried)
Gonimbrasia belina larva

true in Thailand and Laos, where glands of the male insect are the prime ingredient of spicy *mangda* sauces. Other insects popular as food include "witchetty grubs," huge, root-boring larvae of the cossid moth *Endoxyla leucomochla* in Australia, and "mopane worms," caterpillars of the giant silkmoth *Gonimbrasia belina*, found abundantly in southern Africa. Grubs of palm weevils, *Rhynchophorus phoenicis*, are a traditional food in Angola, and other species in the genus are harvested in tropical Africa, Asia, and Latin America. Most of these insects are palatable because they eat plants and lack toxins used in self-defense.

Even in cultures accustomed to consuming insects, entomophagy accounts for only 5%–10% of the annual animal protein in the human diet. The mentality of Western civilization seems to be that eating insects is a behavior we have overcome, having advanced beyond our hunter-gatherer ancestry to farming and livestock ranching. Entomophagy thus remains, sadly, a novelty reserved for dares and reality television.

Epomis spp. Ground Beetles

Entice something to eat you so *you* can eat *it*. That is the mind-bending strategy of beetle larvae in the genus *Epomis*. Amphibians, especially frogs and toads, are primary vertebrate predators of insects, particularly soft and juicy larvae that do not require chewing. Incredibly, these beetle larvae turn the tables.

Epomis is an Old World genus of ground beetles (Carabidae) with thirty known species, most occurring in Africa. Entomologist Gil Wizen has studied them in the lab, and in the field in Israel, if only to make certain his eyes have not been deceiving him. Both the adult insects and the larval stages appear to be obligate, specialized predators of young frogs and toads. *Epomis* synchronizes its metamorphosis with that of its amphibian prey, such that beetle larvae and "froglets" and "toadlets" are present at the same time.

The beetle larva waits on the edge of a temporary pond or rain-pool and lures a prowling toad or frog. The larva waves its antennae alternately and moves its jaws in the same manner. The motion stimulates the amphibian to approach and attempt to nail the poor insect with its tongue. Faster than a speeding bullet, the larva dodges the mouth-missile and launches its own attack, latching onto the frog's mouth or throat. The double-hooked mandibles hold fast and the beetle larva begins drawing body fluids from its prey immediately. It eventually consumes it entirely.

Adult *Epomis* simply approach a small frog or toad and seize it in their jaws, holding on with all six legs to avoid being flung off. The beetle commences chewing and kills and eats most of its prey. The beetle is a beautiful metallic insect, but perhaps appearances can't overcome its macabre lifestyle.

EPT

The acronym for three orders of aquatic insects that are used by environmental researchers and consultants in diagnosing water quality, mostly in rivers and streams.

E Ephemeroptera, the mayflies. Known for their ephemeral life as adults, the aquatic naiads can take years to mature. Most are "collector-gatherers" feeding on fine particles on the streambed, or "scrapers" that eat algae coating the surface of stones. Some actively swim, while others are "clingers" to objects on the bottom. Adults emerge from the last naiad stage.

P Plecoptera, the stoneflies. As aquatic naiads, they are typically "shredders" that tear chunks of plant matter, or predators on other aquatic invertebrates. Most species cling to the surface of stones. Adults of the largest species are the "salmonflies" familiar to anglers. They emerge from the last instar of the naiad stage.

T Trichoptera, the caddisflies. They go through complete metamorphosis. Larval lifestyles are diverse.

Some spin nets or purselike bags, or tubes, and are "filterers," or predators, straining swift currents for very small particles of decomposing organic matter, or tiny aquatic invertebrates. Many species build cases around themselves, using specific types of vegetation or fragments thereof, or pebbles, or sand grains. These casemakers can be scrapers, collector-gatherers, or even predators. Some occur on the streambed or lake bottom, others climb aquatic vegetation and debris. Adults resemble moths.

All three orders are more sensitive to pollutants than other macroinvertebrates, and relatively easy to identify. In addition to chemical sensitivity, they may have limited tolerance to acidity, turbidity, and warmer water temperatures. The EPT index provides a measure of species richness that in turn reflects the overall health of the aquatic habitat from which samples were taken. The less diversity, the more likely pollutants are exerting harmful effects.

See also Ecosystem Services; Exuviae.

Exploding Ants

Even the insect world has its suicide bombers. Certain carpenter ants in the "*C. saundersi* complex" of the genus *Colobopsis* (*C. cylindrica*, *C. saundersi*, *C. explodens*, *C. badia*, *C. corallina*, and probably others) of southeast Asia practice autothysis.

That is, they commit suicide during individual confrontations and battles with other ants, especially weaver ants. Powerful contractions of muscles in the abdomen force the contents of their mandibular gland reservoirs to expel tacky and irritating compounds so violently that the abdomen explodes. It also tears the mandibular gland itself, spraying the goo in all directions from the front of the head. The substance is white, cream, or yellow, and aromatic, with a spicy quality reminiscent of curry. It is quite the cocktail, full of phenols and terpenoids in particular.

This defense is a last resort, of course. *Colobopsis*, known informally as "janitor ants," nest primarily in hollow twigs and stems. The major workers have blocky heads and flattened faces they use to plug the small round entrance holes. Entry by colony members is by the ant equivalent of a keypad, the proper "code" of antenna strokes to the guard's face eliciting an open door as the sentry backs inside.

The term autothysis was coined in a 1974 publication by Ulrich and Eleonore Maschwitz. This kamikaze-like phenomenon is not restricted to ants, either. The termites *Neocapritermes taracua*, found in French Guiana, accumulate stores of copper-containing proteins throughout their lives. Older worker termites volunteer as living bombs should the colony be attacked. Saliva from the labial gland reacts with the crystals upon autothysis, yielding a toxin deadly to other termites. Termites in six other genera self-detonate to block tunnels in their nests, preventing enemies (usually marauding ants) from making further inroads.

Exuviae

When an insect molts, the discarded exoskeleton it leaves behind is called an exuviae. The word is both singular and plural. These ghostly objects often perplex non-entomologists.

Exuvia of a
dragonfly

Two circumstances amplify the mystery of exuviae. They often appear in large numbers in a localized area. Secondly, the immature insects depicted by the exuviae bear little or no resemblance to the adults. This lack of resemblance is especially true of insects that live as nymphs in the soil, or naiads in water, hidden from view until they emerge as winged adults. Molting usually takes place at night, so association between the adult and immature is rarely made.

Aquatic insects like dragonflies, stoneflies, and mayflies leave exuviae clinging to rocks or emergent vegetation. Entomologists can often deduce the species, even the sex, from those vestiges. Mayflies molt twice once they leave the water. The first results in a subimago, a pre-adult that anglers call a "dun." This subimago molts again into a full-fledged adult, the fly-fisherman's "spinner."

A newly minted adult insect is called a teneral. Generally, it means the adult is still soft, with pigments not fully expressed. The term is used differently for dragonflies and damselflies, whereby the young adult insect, even when hardened and a capable flier, possesses coloring different from a mature adult.

Examine an exuviae closely and you may see stringy white filaments protruding from the hollow inside. These are remnants of the major tracheal pathways, representing invaginations of the cuticle that form those breathing passages.

Keen-eyed predators like birds notice clues like leaf damage and insect waste, so most caterpillars consume their exuviae after molting to hide their presence. Other insects detach their shed "skin" and let it fall or be blown by the wind, eliminating all traces of their continued existence on the plant.

See also Metamorphosis.

Fabre, Jean-Henri (1823–1915)

Frenchman Jean-Henri Casimir Fabre is celebrated as the "father of entomology," but it was his ability to captivate nonscientists with his narratives of the life histories of insects that is most remembered and revered.

Fabre was in many ways the archetypical entomologist in his solitude, indefatigable curiosity, sharp and patient observational skills, and innovative thought processes. His career path was never straightforward nor ascending, and he struggled with financial poverty throughout his life. Disdain for the confining rules of academic, social, and political life defined him, and limited his ability to excel via "normal" avenues. His biggest break came courtesy of his friendship with Victor Duruy, Minister of Public Instruction under Napoleon III. In 1870, Duruy invited Fabre, by then 47 years old, to present a lecture series in Avignon, the audience composed in part of secondary school girls. The opportunity to engage young, impressionable minds changed the direction of Fabre's career and led him to advocate openly for teaching science to young women.

Fabre's writings are his most enduring legacy, especially his ten-part *Souvenirs entomologiques* (1879–1909). The texts included prose and poetry related to insects and plants, as well as topics of instinct and heredity, even veering into ethics and anthropological theories. More popular works each detailed the lives of glow-worms, mason bees, flies, hunting wasps, and other arthropods.

One might assume Fabre traveled often, and far, but his entire adult life was centered within a twenty-mile radius of Avignon, Carpentras, and Orange in the south of France. Fabre's life is a testament to the power of a sense of place, and devotion to passionate observation and lifelong learning. To see the world through his eyes, watch the documentary *Microcosmos* (1996), a fitting tribute to him.

See also Pine Processionary Caterpillars.

Fairyflies

Would you believe that certain wasps lay claim to the title of world's smallest insect? The "fairyflies" of the family Mymaridae are stripped-down DNA delivery vessels, the smallest, it is claimed, able to fly through the eye of a needle.

Mymarids are all parasitoids of the eggs of other insects. That is, they are parasites that ultimately kill their hosts. They get to their hosts quickly, before embryogenesis has progressed to any great degree, and their oviposition into the host egg usually halts further development. The mymarid larva apparently lacks

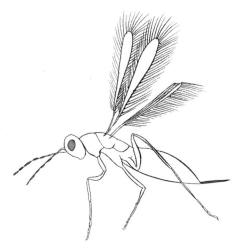

a tracheal system and spiracles, the typical breathing system for insects. It progresses through its own metamorphosis entirely within the host egg.

As delicate as they appear, fairyflies are remarkably durable. Some species have the wings greatly reduced (brachypterous), or lacking completely (apterous). A few are aquatic, using paddle-like wings to row through water. Those species are parasitoids of eggs of aquatic insects such as diving beetles, so they must submerge to locate their hosts. They can carry out mating and host-seeking without leaving the water but may climb emergent vegetation to break the surface film and fly to another pond.

On dry land, fairyflies typically seek insect eggs in concealed situations to avoid competition from other egg parasitoids, namely wasps in the family Trichogrammatidae. Consequently, mymarids search plant tissues in which host eggs can be embedded, or look in flower bracts or between bud scales, or in the soil.

These tiny wasps are among our most helpful allies in the war on pest insects. *Anaphes nitens* is employed to control the weevil *Gonipterus scutellatus*, a pest of ornamental eucalyptus trees in South Africa, New Zealand, South America, and Europe. Various species of *Anagrus* are utilized as biocontrol agents for various leafhopper pests, or are being actively evaluated as such.

See also Biocontrol; Integrated Pest Management.

Fig Wasps

Figs (*Ficus* spp.) do not produce fruit in the truest sense. What we call a fig fruit is a syconium. As keystone species, figs are at the center of many food webs, and their pollination hinges on a mutualistic relationship with certain wasps.

Flowers are located inside the syconium, accessible only to female wasps of the family Agaonidae through the ostiole, a tiny hole at the apex of the fig. It is a tight squeeze even for wasps that average less than 2 millimeters. Once inside, each female wasp goes about laying a single egg in the ovary of each flower. The styles of the florets vary in length, so sometimes she cannot reach the ovary. She still accomplishes pollination in the attempt. A larva hatches from the egg and feeds inside a gall that develops in the floret. At the completion of metamorphosis, which takes three to twenty weeks depending on the fig species, an adult wasp emerges.

Females are wasplike, but adult males resemble predatory beetle larvae. Some individuals possess formidable jaws they use to battle other males for the right to mate with females. After mating, the male bores a passage through the fig wall to allow the female's escape. Before her exit, she harvests pollen from male flowers, or passively accumulates it on her way out. She finds another fig by following an aromatic odor trail broadcast by the fig.

There are many exceptions to the preceding. About half of fig species have male and female syconia on separate trees. The wasps successfully complete their life cycle in the male figs, where all florets allow for successful oviposition. To the female wasps, female figs look and smell identical to the male figs, but inside, *none* of the female florets allow successful oviposition. The female wasps that fall for this trap succeed in pollination but fail to reproduce.

See also Ecosystem Services; Pollinators.

Fire Bugs

A surprising number of pyrophilous (fire-loving) insects flock to forest fires and other smoldering situations. Beetles, flies, some moths, and a few true bugs, collectively representing at least twenty-five families, are known to be attracted to fire.

Beetles have developed the most sophisticated sensory systems for locating conflagrations. The Black Fire Beetle, *Melanophila acuminata*, in the family Buprestidae, is the most studied. The beetle has two pit organs on the underside of its thorax. Inside each pit are 50–100 sensilla that each detect infrared radiation in a range corresponding to the 435°–1150° Centigrade temperature at which forest fires burn. Remarkably, the single dendrite in each sensillum converts the radiation input into a micromechanical stimulus that is then measured by a mechanoreceptor. The beetle essentially reads the radiation intensity as a vibration.

Meanwhile, the antennae of *M. acuminata* are tuned to the volatile chemicals in smoke. This is the longer-distance fire detection system, whereas the heat receptors are effective at impressive distances by themselves, at least 1–5 kilometers. Since buprestids usually attack freshly killed or severely weakened trees, arriving before the competition is of paramount importance. Buprestid larvae, called flathead borers, typically tunnel under bark and/or through the wood itself.

Merimna atrata, a buprestid found in Australia, possesses two pairs of infrared receptors on the underside of the abdomen, but they are unable to detect thermal radiation at great distances. Instead, the receptors help the beetle avoid scorching its feet upon landing, detecting "hot spots" not otherwise perceptible.

Some flat bark bugs in the genus *Aradus* detect forest fires using about a dozen infrared receptors similar to *Melanophila*, distributed across the underside of the front of the thorax. These insects feed on fungi that grow on burned wood.

Flea Circus

How unlikely is it that fleas, one of the worst insect enemies of mankind, should be candidates for our amusement? It is a testament to the entrepreneurial spirit of our species that we can find entertainment in such a ridiculous source.

The origin of flea circuses is murky, but they may have existed in Europe as early as the sixteenth century, according to secondhand information from English writer Thomas Muffet. They perhaps reached their zenith when Signor Bertolotto's "Extraordinary Exhibition of the Industrious Fleas" catapulted the flea circus into the spotlight in the 1830s. Intricate miniature replicas of chariots, carriages, even a Man of War ship, all pulled by "trained" fleas, combined with the convincing salesmanship of Bertolotto, made the theatrics quite the attraction, at only one shilling for admission.

The Human Flea, *Pulex irritans*, was the species of choice in these enterprises, perhaps because they fed readily on the blood of their captors. Bertolotto liked to profess that his charges were fed on "ladies of distinction," and indeed, Human Fleas are attracted disproportionately to women, somehow sensing ovarian hormones.

As hygiene and sanitation practices grew more effective, fleas became scarce, and so did flea circuses. They persist today, but irregularly. In the late 1980s, the century-old Munich Flea Circus, by then run by Hans Mathes, was still alive in Germany. His fleas were

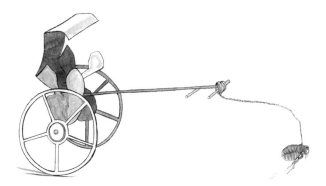

purported to dance, juggle, pull carriages, and spin a Ferris wheel.

Predictably, fake flea circuses arose to compete with, or replace, actual insect-driven acts. These faux flea endeavors capitalized on the fact that it is difficult for spectators to discern fleas in the first place. Whether any young flea ever ran away to join a flea circus remains a mystery.

Florissant Fossil Beds

Amber gets all the glory when it comes to insect fossilization, but some of the most spectacular examples of paleoentomology come from rocks and stones. One of the most prolific sites for insect fossils is Florissant Fossil Beds National Monument in Florissant, Colorado.

The monument was established in 1969, but it represents a far more ancient time, roughly 34 million years ago, preserved in shale. The Florissant Formation covers part of the Eocene epoch, in the Paleogene period of the Cenozoic era. The park is named for ancient Lake Florissant, and preserves about one-third of that now extinct body of water. A volcanic mudflow created the lake by damming streams and building up sediments on the lake bottom. Subsequent eruptions of surrounding volcanoes rained ash and pumice and added more mudflows and stream-deposited sediments. The ash also triggered a proliferation of diatoms in the lake. The result was a series of thin layers of material, diatoms included, that trapped plants, insects, and other organisms between them.

Among the more unexpected fossils is a tsetse fly, *Glossina oligocena*, larger than its modern descendants, which today occur only in Africa. It is one of about 1,500 species of insects and spiders discovered from Lake Florissant. All the specimens represent either compression fossils, impression fossils, or a combination of both. A compression is the actual insect, with a mineralized exoskeleton. An impression fossil is like a footprint: the organism is long gone, but its impression remains.

Fossil collecting within the monument is understandably prohibited, but a private family enterprise does allow digging. They have supplied many fossils to museums and the National Park Service, too.

See also Amber; Tsetse Flies.

Fluorescence

Not to be confused with insects and other arthropods that produce their own light through bioluminescence, other organisms fluoresce when exposed to ultraviolet (UV) light. Many scientists exploit this phenomenon in order to locate and collect specimens. Fluorescence is produced when light of a short wavelength (high energy) is absorbed, then rebroadcast as a longer wavelength (of lower energy).

The best-known example of arthropod fluorescence is in scorpions. It has been determined that the outer layer of a scorpion's exoskeleton, the epicuticle, is where the fluorescence originates. Among the compounds in this scorpion armor is a type of coumarin. This chemical is also found in plants, helping prevent sunburn in vulnerable seedlings. It stands to reason that scorpions, which thrive in arid habitats, would need sunscreen, even if they are mostly nocturnal. In insects, it is more complicated.

As early as 1924, fluorescent pigments were known in butterflies, with scientific papers on the subject by the 1950s. There has been a resurgence in interest since 2001, yet it remains a neglected field of study. As of today, fluorescence is also known in caterpillars, beetles, ants, many butterflies, at least one grasshopper, and one dragonfly. This is likely the tip of the iceberg; and we know only a handful of the compounds linked to insect fluorescence.

Not all insects have fluorescent pigments uniformly distributed over their bodies; and there may be sex differences, whereby males fluoresce and females do not, or vice versa. Complicating matters, UV *reflectance* is different from fluorescence. Sulphur butterflies in the genus *Colias*, for example, are known to use differences in UV reflectance as one mate recognition signal.

See also Bioluminescence.

Frass

Insects, especially in their immature stages, consume great quantities of food and consequently excrete an inordinate amount of waste. The solid feces of insects are called "frass." The term is usually reserved for the fibrous or powdery excretions of wood-boring insects like beetle larvae and termites. Occasionally, frass is broadened to include material discarded by an insect, regardless of whether it has passed through the digestive tract.

Ironically, "frass" is German for "animal feed," so its use as a term for animal droppings is rather inappropriate. Nevertheless, the English definition of frass as insect fecal matter has been in use since at least the mid-1800s.

Insect excretion must conserve water yet lubricate the passage of solid waste. The rectum is so efficient at this that little water is lost and the resulting wastes dry quickly. This is important in circumstances where the insect cannot distance itself from its waste, such as in the tunnels of wood-boring larvae. Wet excrement attracts bacteria, fungi, and other organisms that could threaten infection of the living insect.

Not all waste goes to . . . waste. Some caterpillars, like the Octagonal Casemaker, *Coleophora octagonella*, build portable shelters using their dried feces. Leaf beetles exhibit the most ingenious examples of frass recycling. The larvae of many chrysomelids fashion their excreta into protective accessories. Case-bearing leaf beetles create capsules of hardened feces around their soft bodies. Tortoise beetles and their kin pile their feces onto tail-like spines to form umbrella-like structures that protect them from solar radiation, camouflage them from enemies, or actively deter predators.

Adult warty leaf beetles take a different tack. They are near perfect mimics of the frass of caterpillars, even feigning death should they be discovered by a potential adversary.

Galls

Those strange swellings, bizarre "fruits," and other anomalies on plants are called galls, growths initiated by other organisms, including insects, mites, fungi, and bacteria. The technical term for a plant gall is a cecidium, an innocuous object rarely detrimental to the health of the host.

Among insects that induce galls, gall wasps in the family Cynipidae are most familiar and among the most diverse. Gall midges, family Cecidomyiidae, also account for many galls, though they tend to be less conspicuous. Several families of other flies, plus moths, aphids and their kin, beetles, sawflies, and thrips also include gall-forming species.

How galls are created varies with each gall-former, but they all target plant tissues that are still growing. In cynipids, the initial feeding activities of the larva stimulates growth of the gall. In some sawflies (Tenthredinidae), the adult female wasp applies chemicals at the time she lays her egg, and those substances start the process. Every gall-maker generates unique galls, but how and why is an enduring mystery.

Galls are composed of undifferentiated parenchyma tissues, though in some cases a few cells are specialized to produce structures of the gall. Galls are richer in amino acids, minerals, and other nutrients than surrounding tissue. Many are dense, like a nutshell, or spiny, or otherwise architected to protect the tender occupant(s). Galls are thus food and shelter for the creator.

Cecidia are ecosystems unto themselves. The immature stages of gall-makers are beset by parasitoid wasps, eaten by beetles, birds, mice, and other predators, or share their gall with uninvited guests called inquilines that also feed on the gall tissue.

Humans have used galls for centuries. Oak galls, rich in tannins, have been used since at least the fifth century as a major ingredient of inks.

See also Kinsey, Alfred C.

Giant Water Bugs

If there was an insect identification hotline, one of the most asked-about creatures would be giant water bugs in the family Belostomatidae. They are so notorious they have aliases like "toe-biter" and "electric light bug." Most people see them away from water, which may account for the confusion. The adults fly well but get stranded on land near lights at night.

Belostomatids are predatory, clinging motionless to aquatic vegetation, waiting in ambush. They seize other animals in their beefy, raptorial forelegs, then paralyze the victim with a bite delivered through a short, stout rostrum. The chemical cocktail they inject includes

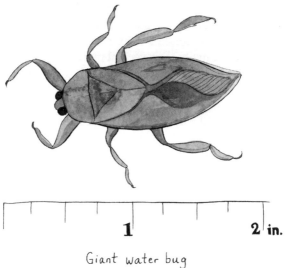

Giant water bug
Lethocerus americanus

powerful enzymes that begin digesting the internal tissues of their catch. The insect then draws out a liquified meal. So formidable are belostomatids that small fish, tadpoles, frogs, newts, salamanders, and even snakes fall prey. Some species in the genus *Lethocerus* can exceed 12 centimeters (4½ inches) in length.

Giant water bugs are not without their redeeming qualities. Male members of the genera *Belostoma*, *Abedus*, *Appasus*, *Diplonychus*, *Hydrocyrius*, *Limnogeton*, and *Weberiella* allow their mates to cement the eggs to their backs. The male not only protects the eggs by default, he periodically suns them and aerates them, too. Such patriarchal brooding behavior is rare among insects. Male *Lethocerus* attend eggs the female lays above the water line on emergent vegetation and other objects. He may moisten the eggs, shade them, and of course guard them from predators. He may nurture more than one clutch, presumably from several mates.

As with sharks, giant water bugs have more to fear from us than we have to fear from them. In parts of southeast Asia, the insects are a food delicacy. Harvesting has reached such a scale that the species *Kirkaldyia deyrolli* is threatened in South Korea and Japan.

See also Entomophagy; Parental Care.

Grylloblattids (rock crawlers)

One major challenge insects face is staying warm enough to be active. "Rock crawlers" or "ice crawlers" in the order Grylloblattodea face the opposite problem. They survive in a narrow range of temperatures, above or below which they die. Rumor has it that the warmth of a human hand can kill them.

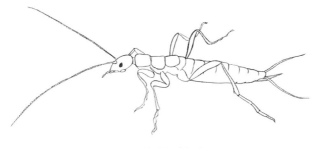

Grylloblattid
Grylloblatta sp.

Grylloblattids are considered relics that were never speciose during their entire evolutionary history. Fossils indicate that precursors to rock crawlers had functional wings that were subsequently lost to them in the Cretaceous or early Tertiary period. Differing results from comparison of physical characters, and molecular DNA analyses, offer conflicting pictures of the relationship of grylloblattids to grasshoppers and crickets or, alternatively, to earwigs. Present-day rock crawlers were discovered in 1913, and today there are thirty-two known species.

As omnivores that scavenge dead or dying insects on the surface of the snow, their preferred temperature range is between 1 and 4 degrees Centigrade. They are presumably nocturnal, living beneath stones, leaf litter, and rocks at the edges of snow fields and glaciers. Blind species exist in certain caves in Asia, while other species exist in lava tubes in arid environments of western North America.

One species, *Grylloblatta chirurgica*, recolonized the blast zone around Mt. St. Helens in Washington state only four years after the major eruption in May 1980. Their lifestyle lends them a long life. It takes six or seven years for the average rock crawler to reach adulthood, and in the laboratory, specimens have lived up to ten years.

Will grylloblattids continue to persist in a warming climate that threatens most of their geographic strongholds? Human encroachment at lower altitudes could also wipe out the few species surviving along rocky streams and in forested areas.

Gynandromorph

In an era when we are finally recognizing and respecting human beings who identify as nonbinary, it helps to remember that in nature, male and female exist in a variety of ways, and not always separately. One striking example is that of gynandromorphs: individuals that possess both male and female characters.

Because of often extreme sexual dimorphism in the insect world, gynandromorphs can be spectacular specimens. Gynandromorphy can be expressed in a bilateral fashion, one side of the organism being female, the opposite side male. Alternatively, it can manifest as a "mosaic," with male characteristics displaying in otherwise female body parts, and vice versa. It is in butterflies that gynandromorphs are most obvious and extreme, but the phenomenon has been observed in many insects, crustaceans and other arthropods, and birds.

The mechanisms that cause a gynandromorph are complex and varied. The most common are related to abnormal aspects of cell division in the embryo, whereby too many, or too few, male or female sex chromosomes end up in different cells. Occasionally, an insect egg can have two nuclei instead of one, and if they are each destined to be a different sex, and both are fertilized, then a gynandromorph may eventually result. Additional factors that can rarely induce a gynandromorph include hybridization, a bacterial or viral infection, mutations, even variations in temperature during egg development.

Gynandromorphs are rarely capable of reproducing, but they are a reminder that maintaining genetic diversity is an overriding priority in nature. Gynandromorphs should perhaps be viewed not as "failures," but celebrated as spectacular expressions of that diversity and the many cellular processes that lead to it.

Gypsy Moth (now LD Moth)

The poster child for the law of unintended consequences may be the LD Moth. Native to Eurasia, it was introduced to Medford, Massachusetts, USA, by Étienne Léopold Trouvelot of France in 1869. Trouvelot hoped to start a sericulture industry there, and was led to believe the LD Moth was a worthy candidate. He should have solicited a second opinion.

Lymantria dispar received its original name of Gypsy Moth in 1742, from Englishman Benjamin Wilkes. This should have been a clue to Trouvelot as to how capable this moth is of dispersing. Alas, the moths quickly escaped control, and by 1889 the entire state of Massachusetts faced a leaf-eating menace. So dramatic was the damage that in 1890 the legislature passed a bill to launch an eradication campaign, earmarking $25,000 for the effort.

Today, the Gypsy Moth, renamed LD Moth in 2021 to avoid further stigmatizing the Romani people of Europe, occurs over most of the eastern United States and adjacent Canada. The species does not disperse itself as you might imagine. Adult female moths are winged, but flightless. They attract mates by emitting a pheromone, and the males can fly great distances to reach them. Once mated, the female lays an egg mass which she covers in scales from her body. The tiny caterpillars that hatch each issue long strands of silk from glands in their mouthparts. The filaments catch in the wind and the caterpillars are blown far and wide.

LD Moths get assistance in their travels when humans inadvertently transport them. Egg masses can be adhered to the wheels of vehicles, trailers, and storage pods. This is only one aspect of their biology that makes them pestiferous. While many insects are restricted in their diet to a few kinds of plants, more than 300 species are on the menu of the LD Moth.

See also Riley, Charles Valentine; Sericulture.

Hair-Pencils

Pheromones are not always a product of female animals, used to signal males as to their sexual receptivity. Many *male* insects use pheromones to communicate their fitness as potential mates. Moths, and some butterflies, deploy those pheromones by everting glands and/or bundles of hairs that emit those volatile compounds.

In 2017, a video of a particularly extravagant example went viral on social media. A male tiger moth, *Creatonotos gangis* from southeast Asia and Australia, was shown with what looked like half of a hairy octopus coming out of it's a**. Such glands are called coremata, a Greek word that translates roughly to "feather dusters," an apt description of most of these organs.

The scents broadcast from those rear-end accessories are derived from plant chemicals consumed in the larval stage, or sometimes the adult stage. Males of certain milkweed butterflies (Danainae) visit wilted and

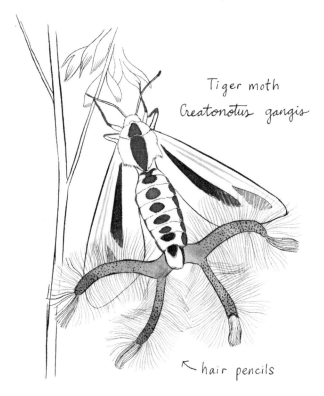

Tiger moth
Creatonotus gangis

← hair pencils

dead leaves of *Heliotropium*, *Tournefortia*, *Crotalaria*, and other plants to imbibe pyrrolizidine alkaloids (PAs), potent compounds that afford the insect protection. From the PAs, the butterflies synthesize dihydropyrrolizines, which they use in their hair-pencils. A male's potential mate infers that the stronger the scent, the greater the quantity of PAs he can transfer during mating.

Our tentacled tiger moth friend, *Creatonotos*, can sequester plant compounds only in the caterpillar stage. Depending on the quantity of chemicals consumed, a given male caterpillar may not metamorphose into a well-endowed adult like the one in the video. He could still benefit by association with other males, though. They tend to congregate to display for females.

In the case of the milkweed butterflies, male "equipment" is deployed in flashes of his bushy tushy while he hovers over a prospective mate, sometimes contacting her antennae and face in the process.

See also Pheromones.

Hellgrammites
Few insects are as vexing to non-entomologists as dobsonflies. Seldom do people associate the aquatic larvae with the terrestrial adults, nor do they recognize the male and female adults as the same species. Conspicuous as they are, much of their biology remains a mystery.

Precious little informs the etymology of "hellgrammite," so we are left to speculate. "Hellgrammite," "dobson," and "crawler" are, historically, names fishermen applied to the larvae. Perhaps it is a compound creation of "hell" plus "grima," "grimman," or related Old English that would translate loosely to "goblin from hell."

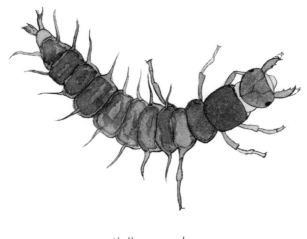

Hellgrammite
Eastern dobsonfly
Corydalus cornutus

Fishermen frequently use the huge (nearly 9 centimeters) larvae as bait.

They are impressive river and stream animals that capture prey that drifts by on the current. Each segment of the abdomen is adorned with a pair of fleshy, fingerlike filaments. The purpose of these extensions is unknown. Two anal appendages at the rear are equipped with claws to anchor the insect to the streambed. The larva stage may last two or three years in northern latitudes. The hellgrammite crawls onto land to pupate, often under stones or other objects. The pupa is active, with free appendages capable of propelling it if necessary.

Adult male dobsonflies have long, curving mandibles reminiscent of ice tongs. A well-endowed male may measure more than 8 centimeters. Anecdotal evidence suggests they use these weapons to battle other males. Females have typical jaws, like the larvae, and bite hard if molested. Females paste masses of about 1,000 eggs onto vertical substrates overhanging water, covering them with a clear liquid that dries into a white, chalky substance. They typically guard the mass for a short period. The eggs hatch in about one week, and the larvae drop or crawl into the water.

Hexapods, Non-insect

Surprise! Not all six-legged creatures are insects. The most primitive, in the evolutionary sense, are "non-insect hexapods" in three classes by themselves: Protura, Diplura, and Collembola. Some authorities still include all three in Insecta, but they are currently a minority opinion.

Non-insect hexapods have in common a primitive head structure whereby the mouthparts are hidden within a pouchlike orifice. This arrangement is termed entognathous. Additional characteristics of the Entognatha are reduced, vestigial, or absent compound eyes and, internally, reduced Malpighian tubules. The exoskeleton is typically soft. Fertilization of eggs is external.

Proturans are *de facto* tetrapods, as the front legs, covered in sensory receptors, function as antennae. About 500 species of proturans are collectively distributed around the globe, but they are minute (<2 mm), dwell in soil, leaf litter, moss, and rotting wood, and are seldom encountered. They presumably feed on mycorrhizal fungi.

Diplurans average 7–10 millimeters and occupy niches similar to those of proturans. One subgroup of diplurans is vegetarian, the other predatory on small invertebrates. Predatory species possess stout forceps at the rear with which they seize victims. There are roughly 800 species of Diplura, collectively found worldwide.

Anyone familiar with the abundance and diversity (about 6,000 species) of springtails, class Collembola, understands that "primitive" does not equal "unsuccessful." Springtails occur almost everywhere, including the surface of puddles and ponds, on snow, even indoors (check the soil of potted plants, or look closely in the bathtub). Springtails are unique among hexapods in that they continue molting after reaching sexual maturity. They are vital members of soil fauna, feeding mostly on decaying organic matter and associated fungi.

Ancestral prototypes of these hexapods likely emerged sometime in the Devonian or late Silurian period, about 400 million years ago.

See also Snow Insects.

Hilltopping

Humans may be inspired to climb a mountain "because it's there," according to English adventurer George Mallory, but some insects have a different purpose in flying to summits. Some male butterflies, wasps, ants, beetles, dragonflies, and flies head for ridgetops, buttes, and other promontories as rendezvous sites for mates, a behavior known as "hilltopping."

This strategy evolved in species where females are dispersed, not concentrated at emergence sites, nesting sites, or resources such as flower patches or water. It may involve the individual male defending a territory that he patrols, or overlooks from a perch. It is exhaustive, as he constantly engages rivals. Battles are mostly ritualistic, vertical or horizontal pursuits, but some butterflies have violent collisions.

More typically, nonterritorial patrolling or perching is the norm, in which case males engage in "scramble competition" for any female that passes by. Both aggressive and passive styles of hilltopping are defined as lek polygyny. A lek is a location where males congregate to attract females. A common example among birds is the sage grouse, where groups of males display to females. Polygyny suggests males having more than one mate simultaneously, but for hilltopping insects it means males mate with several females in succession if opportunities allow.

Virgin females are more likely to frequent hilltops than mated females, another benefit of the male strategy. Mated females usually go about seeking host plants, or begin nesting, activities that invariably lead them away from mating sites.

We still know little about which insect species habitually engage in hilltopping, let alone what particular strategies males use for maximum success. Some appear to favor the tallest *object* on a hilltop, such as a tree or bush, in what amounts to landmark lekking.

Honeydew

The liquid waste products of aphids, scale insects, treehoppers, and other true bugs is addictive to a variety of other insects. Called honeydew, it is sometimes preferred over flower nectar for a carbohydrate boost.

Park your car under an aphid-infested tree and you will become intimately acquainted with honeydew. The substance is sticky and accumulates dust, pollen, and other particles. Sooty molds quickly colonize the residue on foliage, compromising the ability of the plant to photosynthesize. To avoid drowning in their own waste, aphids may kick droplets with their hind legs. Other aphids are coated in waxy secretions that repel honeydew. "Sharpshooter" leafhoppers expel jets of honeydew some distance from their bodies. Nymphs of certain psyllids mix honeydew and waxy secretions to form a protective, crystal-like coating called "lerp."

Honeydew is produced by insects that suck phloem sap. Phloem is not terribly nutritious, so it passes largely unprocessed through the creature. Besides representing up to 90% of the sugars taken in by the insect, honeydew contains metabolic waste products, and a few gut bacteria.

Ants crave honeydew to the degree that some species have symbiotic relationships with aphids, scale insects, or planthoppers. The ants "farm" their flock of food-producers, soliciting drops of honeydew by stroking the rear end of the insect. Some ants build protective "barns" over aphid colonies, and all vigorously defend their charges from predators and parasitoids.

Other insects depend on honeydew for nourishment when flower nectar is scarce, particularly in early summer and late autumn. Birds in tropical Mexico and Australia that feed on flower nectar also partake of honeydew. Flying foxes of Australia regularly feed on lerp from the Lerp Psyllid. Geckos in Madagascar solicit honeydew from flatid planthoppers, much as ants do with aphids.

Honeypot Ants

One of the most unique examples of insect survival is the food storage strategy of honeypot ants. Specialized worker ants are used as living barrels for holding liquid sweets. The thirty species of *Myrmecocystus* in western North America are the most celebrated of these ants, but certain species in seven other genera exhibit similar behaviors: *Camponotus* and *Melophorus* in arid Australia; *Anoplolepis* in southeastern Africa; *Cataglyphis* in north Africa; *Leptomyrmex* in Melanesia; *Plagiolepis* in the Old World; and *Prenolepis*, cosmopolitan in distribution.

Physogastry is the scientific term for a distended abdomen in insects. In this case it is owed to the elasticity of the ant's crop. The crop is an internal reservoir where liquid food is retained for regurgitation later. In honeypot ants, it can swell the abdomen to the size of a grape. Worker ants dedicated to this task are called "repletes," and they cling to the ceilings of the deepest chambers in the subterranean nest. Repletes can return to their normal dimensions when . . . depleted.

Repletes store honeydew from aphids, scale insects, and certain galls, but the colonies are predatory, too. *Myrmecocystus mimicus* hunt termites, but foraging leads the ants close to adjacent colonies. When foragers from one nest intersect another, a ritualized "tournament" ensues, each colony measuring the strength of the other through posturing. If one colony has fewer "major" workers capable of combat, the stronger colony may raid it, killing the queen and stealing workers, including repletes. Meanwhile, minor workers of the interloping colony continue to capture and kill the termites they were after in the first place.

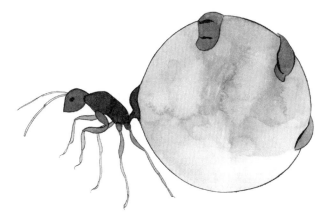

Honeypot ants are a delicacy to many Australian ab-origines, but the women dig them up, a feat of grueling labor. Indigenous peoples of North America also ate honeypot ants.

Horse Guard

Before you swat the wasp hovering around the legs of your favorite steed, take a second look at what it is doing. The North American sand wasp *Stictia carolina* is known informally as the "horse guard," for good reason.

The Horse Guard is solitary, each female excavating her own nest burrow in sandy, well-vegetated soil, ide-ally on a gentle slope. Such locations are few, so many wasps may occupy a small area, though the second generation tends to expand the periphery, with less den-sity. The constant activity of these large insects adjacent to croplands or pastures can be unnerving to humans, but fear not.

Female Horse Guards, after completing a nest burrow, must find food for their larval offspring. Growth demands protein, so the mother wasp goes hunting for flies. Horse flies and deer flies are hefty and fit the bill nicely, especially for older wasp larvae. Females will hover around the legs of livestock and pluck the blood-sucking dipterans as soon as they alight, or even in midair.

Sand wasps of many species, including *Stictia carolina*, practice progressive provisioning. This behavior is in contrast to most solitary wasps that mass-provision, storing a stock of paralyzed prey all at once, then leaving the wasp larva to consume it without supervision. The doting sand wasp mother brings food as needed. These repeat visits to the nest allow her to dispatch any parasitoids or predators that might have snuck into the nest during one of her departures. She may or may not close the nest between trips, but she finds it again, unerringly, provided no one has stepped on, or driven over, the spot. We, on the other hand, can't remember where we parked.

Hummingbird Moths

One way to recruit bird-watchers (they prefer the term "birders" now) to the hobby of . . . bug-watching . . . is through an insect that looks and acts like a bird. Enter certain hawk moths (aka sphinx moths) in the family Sphingidae to bridge that divide. They look nothing like hawks, but they behave much like hummingbirds.

Only *Macroglossum stellatarum* bears the official name "Hummingbird Hawk-moth." It ranges from the Mediterranean region to Japan but is a strong enough flier that it frequently strays farther north and south. The

Hummingbird hawk-moth
Macroglossum stellatarum

name "hummingbird moth" has been misapplied to almost all day-flying sphingids, and even nocturnal species that hover in front of flowers like the namesake birds. Some of these moths are *larger* than hummingbirds.

Hovering allows the moths to visit more flowers in a shorter span of time than alighting on each blossom individually, but there are trade-offs. The insects can overheat or, alternatively, not be warm enough for optimal performance. Cold moths elevate their body temperature by shivering, delivering neural impulses to the huge flight muscles, and causing synchronous contractions instead of commands for alternating upstroke and downstroke. Eventually, this warms the thoracic cavity enough to make flight possible.

Moths in danger of overheating dissipate the excess through the abdomen, which acts like a radiator. The separation of oxygen delivery (tracheal network) from the open circulatory system (hemolymph) helps make this dissipation possible. Sphinx moths can also choose when to be active. They fly at night when daytime temperatures are hot, and fly during the day when night-time temperatures are too low.

Many flowers depend on hawk moths for pollination. These plants typically have white, pale yellow, or pink blossoms, easily detected in the dark. They are also highly fragrant, as hawk moths have keen olfactory senses.

See also Pollinators.

Imaginal Discs

Insect metamorphosis is a magical transformative process, but in some ways it is less discreetly parceled than it appears externally. Enthusiasts of entomology and etymology delight in the term "imaginal discs" as a perfect description of how the imago, the adult insect, is "imagined" early in the life cycle.

A definition of imaginal discs is elusive, as all are not created equal. They appear confined to the higher orders of insects with complete metamorphosis, but are not always easily discernible. Imaginal discs usually become evident in the larva, but in a few species, they originate during embryogenesis. Basically, they are clusters of cells, associated with the epidermis, that are destined to become adult features such as antennae, compound eyes, wings, and reproductive organs. Individual cells of this nature are called histoblasts, common to all insects. Typically, imaginal discs present as invaginations of the epidermis, in contrast to, say, wing pads that develop externally as evaginations in grasshoppers and true bugs.

As is usually the case, much of what we know about insect development comes from studies on the laboratory "fruit fly," *Drosophila melanogaster*. What applies to other insect species is speculative, but the following is generally inferred. The invaginated tissues of an imaginal disc form pockets inside of which the adult organs begin to take shape. Homeotic genes direct the development of structures in the insect body, usually by encoding transcription factor proteins that in turn influence "downstream" gene networks. Hox genes, a subset of homeotic genes, code for positioning of histoblasts to guarantee that the adult attributes develop in the correct place on the body. Considering all that can go wrong, it is a minor miracle that metamorphosis usually results in a perfect adult insect.

See also Juvenile Hormone; Metamorphosis.

Insect Apocalypse

Many scientists assert that evidence of a precipitous decline in insect abundance and diversity is circumstantial, but that does not mean that an "insect Armageddon" is not upon us, or in the offing. As the Xerces Society puts it, we know enough to act now.

It is telling that the first alarm sounded from Europe, where the historical loss of megafauna like wolves and bison has created keen awareness of the vulnerability of remaining species. The Krefeld Entomological Society in Germany, an organization of respected amateurs, reported in late 2017 that surveys they conducted in sixty-three nature preserves, over the course of thirty years, revealed drastic declines in insect abundance. This captured global media attention and initiated a backlash from other corners of the entomological community. There exist few comparable quantitative assessments over time, resources having been channeled into economic entomology related to human health, agriculture, and forestry, instead of evaluations of ecosystem health.

Observations in Australia and Puerto Rico offer anecdotal support for a picture of insect decline, but ornithologists in Canada, studying the Whip-poor-will, offer compelling evidence. These insectivorous birds are eating fewer large insects than they once did. Smaller insects accumulate different forms of nitrogen than larger insects, and this elemental "fingerprint" transfers to predators. Analysis of the birds, including museum specimens dating back decades, thus illuminate the change.

Widespread agreement *does* exist for the major causes of biodiversity declines: habitat destruction, pesticide use, and climate change. How to avoid despair? Do not underestimate the ability of natural systems to rebound from even catastrophic events. Assert your right to a healthy planet with its full complement of species. Turn your lawn into a meadow or prairie. Support biodiversity initiatives, and call for a rejuvenation of field biology to establish baseline data.

See also Endangered Insects; Xerces Society.

Insect Fear Film Festival

Halloween has its horror movie marathons, but each February, at the University of Illinois in Urbana-Champaign, big bug blockbusters are celebrated. The film fest has been going strong since 1984, thanks to

Dr. May Berenbaum, a fellow of the American Academy of Arts and Sciences (1996), and recipient of the National Medal of Science (2014).

Dr. Berenbaum was inspired by a showing of the original *Godzilla* by the Asian American club while she was a grad student at Cornell in the late 1970s. The entomology faculty at Cornell thought such cinematic fantasy was beneath their profession and squashed the idea. Still, she persisted, and proposed the festival to her department head at U of I in 1980. Berenbaum and her students see the event as an opportunity to inform as well as entertain, introducing each feature with an explanation of why insects the size of buses are an impossibility, for example.

It is interesting to note that nearly every insect horror film revolves around a storyline in which human interference with the natural world results in calamity. Giant insects are essentially the punishment for crimes against nature, such as radiation from atomic bombs, pollution from toxic waste, or genetic engineering designed to bestow insect "powers" to humans. What could go wrong?

The festival appeals to fans of cinema, certainly, but also brings out those curious about insects, and no one goes home disappointed. It has been so successful that it has been covered by major newspapers, magazines, television, National Public Radio, and even heralded by scientific journals. Berenbaum is one of those rare scholars who can address both professional colleagues and neophyte citizens with equal success, and always with humor. Long live the Insect Fear Film Festival!

Integrated Pest Management

Agricultural pest control has been a series of advances, each replacing the previous generation as it becomes less effective. Only recently have strategies been used in combination to achieve desired outcomes.

In the United States the concept of integrated pest management (IPM) arose largely in response to increasing public resistance to widespread use of certain pesticides, as well as resistance to the chemicals by the target insects themselves. The term was coined by Ray F. Smith and Robert van den Bosch in 1967 and accepted by the National Academy of Sciences in 1969. It was in practice as early as 1959, when it became apparent that pesticides were not the panacea that post World War II formulations promised.

The full arsenal of IPM includes cultural controls such as crop rotation and tilling, employing natural enemies of pests, use of growth regulators to disrupt metamorphosis, and physical controls such as trapping. Today, genetically modified organisms (GMOs) have taken center stage, and received the same criticisms, justified or not, that pesticides once did. IPM is not a one-size-fits-all solution, and some techniques are more applicable to enclosed spaces such as greenhouses and silos. Methods developed to address agricultural pests often become templates for home and garden pest control. IPM has, so far, been most successful in cultivation of rice in developing countries of tropical Asia.

Corporate-driven products continue to dominate the global conversation around pest control, with market share and prevention of patent infringement among top priorities. As long as the scale of agriculture remains so grand, and profit the overwhelming incentive, it is unlikely that meaningful reform will take place. The expectation that developing nations are obliged to feed Western civilization first, a remnant of colonial mentality, is another persistent notion that must be challenged.

See also Biocontrol.

Jumping Beans

The more nostalgic reader will recall that once upon a time, Mexican jumping beans were a commercial novelty item you could find at the five-and-dime. What *are* jumping beans, anyway?

The animated objects are seeds of at least two Mexican shrubs in the spurge family Euphorbiaceae. *Sebastiania pavoniana* grows mostly in dry, rocky terrain in the states of Sonora and Chihuahua and is the source of the familiar retail curio. The plant produces pods, each with three seed-containing capsules called carpels. When ripe, the pods burst explosively.

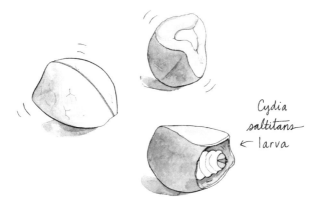

Cydia saltitans
← larva

While the plant is flowering, a small moth, *Cydia saltitans*, visits the blooms and the females lay eggs in the developing ovaries. The caterpillar that hatches enters and eats the carpel, hollowing-out the interior into a chamber it then lines with silk. When these seeds are catapulted from the pod, the larva inside may find itself in a capsule exposed to the scorching sun. Rather than roast, the caterpillar convulses, bouncing its seed-house until it reaches shade.

Eventually, the larva cuts a round hole in the seed wall, but not all the way through. It may lie dormant for months before pupating, but when the adult moth is ready to emerge, the pupa bursts through the "hatch" the caterpillar created.

Tambuti grows as a small tree in arid southern Africa. It produces fruits with three segments, some of which may be entered by the caterpillar of *Emporia melanobasis*. Like the jumping bean, the bouncing fruit segments owe their behavior to the larva inside. "Jumping galls" are produced by the cynipid wasp *Neuroterus saltatorius*. The galls are minute, spherical structures on the underside of white oak leaves that eventually dry, detach, and fall to earth. The wriggling larva inside causes the galls to "jump."

Juvenile Hormone

Stop the presses, insects have found the fabled fountain of youth. Well, not exactly, but their metamorphosis is overwhelmingly moderated by juvenile hormone (JH). It is, quite literally, what keeps them young, for in its reduction or absence the next phase of the life cycle commences. JH is actually a family of hormones, each with slightly different effects, and more than one of which is usually present in an insect.

We owe the discovery and initial knowledge of juvenile hormones to Sir Vincent Wigglesworth's research on insect development in the 1930s, and isolation of the hormone by Carroll Williams in 1956. H. Röller determined the chemical structure of JH in 1967. Juvenile hormones in all insects are the product of paired glands, the *corpora allata*, located posterior to the brain. Juvenile hormones usually cooperate with other hormones called ecdysteroids to coordinate the expression of genes specific to the immature stage just prior to each molt between larval or nymphal instars. At the end of the larval stage in insects with complete metamorphosis, JH levels drop considerably, and the ecdysteroids take over to prime genes that express pupal or adult characteristics. Juvenile hormones reappear in the adult insect to influence reproductive development and function. Polymorphism, caste determination in social insects, and diapause are also regulated in part by juvenile hormones.

So powerful is JH that it has been developed as a weapon against pest species. As early as the mid-1960s, scientists began scheming to synthesize JH for use as "third generation" pesticides. Synthetic juvenile hormone mimics are among a class of compounds known as "growth regulators," but they are expensive to produce and unstable when exposed to light. They cause insects to remain eternally immature, and die before reproducing.

See also Integrated Pest Management; Metamorphosis; Wigglesworth, Vincent Brian.

Kentromorphism

The difference between a grasshopper and a locust is circumstance. A few species of grasshoppers are subject to a unique crowding-induced metamorphosis called kentromorphism. Referred to as "phase change," the default condition is "solitary," while crowding in the nymphal stages produces a "gregarious" phase.

In 1921, in the wake of a locust plague in Stavropol province (Russia) of Northern Caucasus, Sir Boris Usarov recognized that *Locusta migratoria* and *Locusta danica* were not distinct species, but two phases of *L. migratoria*. Favorable weather patterns make possible the localized population explosions that blossom into swarms of billions of individuals. Rains, followed by extended drought, can spark phase change by forcing normally solitary nymphs into concentrated groups in diminishing patches of forage. In the Desert Locust, *Schistocerca gregaria*, the transformation begins within two *hours* of repeated stimulation of the femur segment of the hind leg, from constant contact with conspecifics. This boosts serotonin, a neurotransmitter chemical, which causes the nymphs to seek out the company of others instead of dispersing. A more streamlined, longer-winged adult insect is the culmination of the process.

Eleven species of grasshoppers are currently recognized as having periodic gregarious phases. All occur chiefly in Africa and Eurasia, except the Central American Locust, *Schistocerca piceifrons*. In North America, grasshopper species with occasional outbreaks of epic proportions may exhibit migratory behavior, but they do not undergo any associated physical changes.

Climate change may accelerate the frequency of locust plagues, may render new regions as "outbreak areas" likely to host the initial stages of swarms, permit swarms to flourish for longer periods, or all of the above. Perhaps none of the above. Our understanding of grasshopper population dynamics is in relative infancy. The serotonin connection, for example, was discovered in 2009.

Killer Bees

Before "murder hornets," there were "killer bees," the OGs (Original Gangsters). More like "gangsistas," since colonies are composed of female siblings. In the 1970s, their rage was all the rage, fodder for a recurring skit on *Saturday Night Live*, and several B-movies. Where are they now?

Killer bees are honey bees, *Apis mellifera*. Honey bees are native to the Old World, the most docile subspecies occurring in temperate zones. African varieties are more easily agitated. You would be temperamental, too, if honey badgers were tearing open your hive, and bushmen were smoking you out with flaming elephant dung.

Colonialism brought subspecies of *Apis mellifera* to North America and Latin America in the 1600s. Much cross-breeding ensued, and one African subspecies, *Apis mellifera scutellata*, was targeted for its superior foraging behavior. In 1957, Brazilian bee geneticist Warwick Kerr imported seventy queens of this race to Sao Paulo. Forty-seven survived, but it was the twenty-six that escaped . . .

"Africanized" bees defend their colonies by recruiting more nestmates, pursuing the enemy greater distances, and stinging more readily than "Italian" bees. They also pose threats to native stingless bees, outcompeting them in pollen-and-nectar gathering, and introducing diseases for which stingless bees have no immunity. African bees swarm frequently, rob honey from neighboring hives, and their drones are superior at mating. Consequently, killer bees expand their range in the New World by roughly 200 miles per year.

Colony Collapse Disorder, a mysterious syndrome that has afflicted beekeeping periodically for decades, made headlines in 2006–2008. Suddenly, there were no "bad" bees. In fact, African bees are more resilient to attack by the *Varroa* mite, the arch nemesis of apiculture. Honey bees constitute a major industry, complete with marketing campaigns. Killer bees have been kept out of the spotlight.

Kinsey, Alfred C. (1894–1956)

What does the father of the study of human sexuality have to do with entomology? Prior to his public notoriety, he was best known among scholars for his work on gall wasps of the family Cynipidae.

Kinsey, in a twenty-year span at Indiana University, traveled to thirty-six states and Mexico, logging 18,000 miles, 2,500 of those on foot. He collected 300,000 specimens in the process. Over his career, Kinsey amassed an astounding 7.5 million galls and the wasps that emerged from them. The collection now resides at the American Museum of Natural History. Museum workers are *still* sorting specimens.

He was a data hoarder, the better to understand his subjects. This was astute, as the larger the sample size, the less deviations skewing the results. Data points can inform trends and track changes over time. Still, by his own admission, Kinsey derived power and authority through the accomplishment of quantitative collecting. He also took innumerable measurements of each specimen, recording them in code. His first major paper was "Life Histories of American Cynipidae," published in the *Bulletin of the American Museum of Natural History* on December 20, 1920.

Kinsey began teaching a "marriage course" at IU in 1938, while still working on wasps. One subject he wished to address was human sexuality, but Kinsey found an absence of data on the topic. He adapted his approach to studying gall wasps to studying people in a different habitat: the bedroom. His cold analysis of human behavior was necessary to avoid making interpretations based on morality, religion, and politics. As he had done with his wasps, Kinsey recorded countless measurements and took copious, coded notes. Ultimately, his research included more than 11,000 interviews, and the publication of two books (three more were published posthumously).

See also Galls.

Kissing Bugs

Don't let the bed bugs bite, but by all means avoid "kissing bugs," large (20–44 millimeters) true bugs that feed on vertebrate blood. These unique assassin bugs in the family Reduviidae are also known as blood-sucking conenose, Hualupai tiger, vinchuca, benchuca, and barbeiro. Some species are vectors of a disease, making them villains throughout the tropical Americas.

Assassin bug
Triatoma sanguisuga

These vampire bugs are members of the subfamily Triatominae, with about 18 genera and 138 species collectively found throughout the tropics, plus the southwestern and eastern United States. Five species are major vectors of American trypanosomiasis (Chagas disease) in Mexico, Central America, and South America.

Chagas has been a largely neglected disease since it overwhelmingly afflicts impoverished rural communities where people live in close quarters with livestock and rodents. Increasing numbers of immigrants coming into the United States have raised concerns about

Chagas, but that says more about ethnic intolerance than scientific reality. Native wildlife, especially wood-rats, raccoons, and opossums, are already reservoirs of Chagas. The *Triatoma* species found north of Mexico are simply horrible at transmitting the disease.

The trypanosomes that cause Chagas are found in the feces of the insect. During feeding, the bug may defecate on its human host. Scratching a bite later, the host may unwittingly inoculate themselves. U.S. species are "potty trained" in that they seldom, if ever, defecate while feeding. Cases of dogs contracting Chagas are most often the result of the canine eating one of the insects.

The name "kissing bug" stems from the bug's apparent preference for biting humans on the lips or face while the victim is sleeping. They avoid crawling under blankets and bedding. Absent Chagas, a bite still results in swelling, inflammation, and itching that can last weeks; and some people experience allergic reactions.

Kleptoparasitism

If you have ever had houseguests who live as freeloaders with no regard for the extent of your hospitality, then you can empathize with the victims of kleptoparasitism (or cleptoparasitism).

Food theft behavior may have evolved from an inquiline ("guest") lifestyle, whereby an organism shares the food stores of a host without conflict. A kleptoparasite hogs the entire resource, viewing the host as competitor.

A surprising number of insect kleptoparasites feed on prey trapped in spider webs. Some scorpion flies in the family Panorpidae deftly avoid becoming entangled in the sticky strands to feast on the spider's stored victims. Hover wasps, *Parischnogaster* spp., do exactly that, to pluck prey from spider webs.

Many flies steal food from other insects, and from spiders. Freeloader flies (Milichiidae) feed on the kills of crab spiders, assassin bugs, or robber flies, while the predator itself is eating. Mosquitoes in the genus *Malaya* accost worker ants and force them to regurgitate liquid food. "Highwayman beetles," nitidulid beetles in the genus *Amphotis*, likewise intercept traveling ants and demand a regurgitated "toll."

Cuckoo wasps, velvet ants, and cuckoo bees are obligatory kleptoparasitoids of other solitary wasps and bees. That is, they eventually kill their host. They invade the nests of their hosts and lay an egg in one or several cells. In some species, the adult parasitoid destroys the host egg or larva. If not, the kleoptoparasitoid larva may do so, before it consumes the pollen or insects stored by the host.

Insects may take security measures to defeat kleptoparasitism. Instead of storing a quantity of food all at once, mother sand wasps bring fresh, dead flies to their offspring as needed. Other wasps create fake "accessory burrows" to distract thieving enemies from the real nest.

Lac Insect

That apple or hard candy you are about to eat is probably coated with shellac, a product created by *Kerria lacca*, the Lac Insect. Relax, it is a harmless substance with amazing properties.

Like cochineal, lac is a scale insect that feeds on the sap of its host. The Lac Insect excretes copious amounts of a resinous waste that hardens into a "cell" that covers the insect. The cells of adjacent individuals coalesce into a crust that can accumulate up to half an inch thick. This "sticklac" adhering to twigs and branches is the raw substance from which shellac, dye, and other products are derived.

The production of lac is a major industry in India, where about 75% of the world's supply is generated. Thailand and Vietnam also produce lac. The sticklac cut from host trees and shrubs is scraped and filtered to yield "seedlac," the main ingredient in varnishes, cosmetics, perfumes, and other products. Today, improvements in purification of seedlac have kept this natural product relevant in an era of cheaper, petroleum-based competitors. It is still a preferred finish for many woodworkers and other artisans.

Lacquer is not the only use for lac. Lac dye is composed of anthraquinones that have demonstrated antibiotic, antiviral, and anti-obesity effects. A study from 2016 showed the dye to be promising as a cancer drug. Glazed pills make foul-tasting meds easier to swallow; and since stomach acid has no effect on shellac, prophylactics destined for absorption through the intestines are coated with shellac.

The heyday of shellac was about 1930 through the mid-1940s, but its persistence as a fundamental industrial and agricultural ingredient, and potential for medicinal use, make the Lac Insect one of the most important organisms in our economy.

Lice, Human

These days, we do not think much about lice, unless one of our children comes home with a note from the teacher. It is unfortunate that there is a stigma attached to having lice, that we consider them a sign of filth or moral weakness. Outbreaks of head lice, after all, are often the result of kind children sharing hats and scarves.

Two species of lice are peculiar to humans, and both are "sucking lice" that feed on blood. The human body louse, *Pediculus humanus*, is relatively uncommon and departs the host after feeding. They might better be called "clothes lice," as that is where they normally reside between meals. The head louse, *Pediculus humanus capitis*, restricts itself to the human scalp and neck. The crab louse or pubic louse, *Pthirus pubis*, is adapted to occupation of our nether regions.

Body lice have devastated human populations through transmission of typhus, especially under conditions of war, in prisons, and in other circumstances of overcrowding coupled with lack of opportunities for good hygiene. At least 3 million people died of typhus

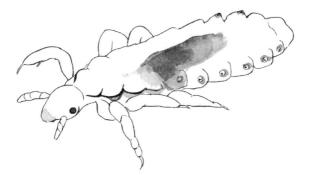

Head louse
Pediculus humanus capitis

between 1917 and 1923 alone. Even World War II saw hundreds of thousands of deaths from typhus before applications of DDT finally stemmed the tide. Typhus is largely in remission today, but there was an outbreak in Russia in 1997, and another in China in 1999.

Head lice are, by comparison, a mere nuisance. In the mid- to late 1990s, they experienced a resurgence, attributed in part to resistance to various insecticides used to treat them, especially pyrethroids; and perhaps also to the unwillingness of parents to apply chemicals to their children. Ironically, remote learning and social distancing related to the coronavirus pandemic may also reduce the incidence of lice outbreaks.

Light Pollution

Artificial lights at night are an underappreciated insect mortality factor, especially for species already suffering from habitat destruction and fragmentation. The effects we know of are significant, but there is much to learn about impacts over time.

Entomologists use blacklights and light traps to attract specimens for collection, but lights are everywhere, in the name of safety and security, advertising, and illumination of around-the-clock workplaces.

Both nocturnal and diurnal insects suffer. Photophobic species may find it impossible to locate dark refuges. Day-active species may extend their periods of activity and suffer shortened life spans, poor behavioral performance, or both. Even the "glow" of a distant metropolis can inhibit the ability of some insects to navigate by celestial patterns. The bioluminescent signals of fireflies are overwhelmed.

One study estimates that one-third of individual insects drawn to a stationary light at night will die before morning, from exhaustion, predation, or the heat of the light itself. Insects may remain immobile and exposed come sunrise, easy pickings for birds and other daylight predators. Insects are distracted from finding mates or engaging in other normal behavior.

Surprising effects of artificial light concern disruption of reproduction and development. Female aquatic insects like mayflies can be confused by the amplified polarized light and attempt to lay eggs on inappropriate surfaces instead of the water. Prolonged exposure to light can result in sterility in males of some insects, and suppress sex pheromone production in females of others.

Astronomers push for "dark sky" initiatives in cities. Entomologists should join them. Meanwhile, turn off unnecessary outdoor lights. Replace continual-emission lights with motion-sensitive fixtures. Trade white lights and blue lights for amber-colored lights. By all means, ditch those bug zappers.

See also Endangered Insects; Insect Apocalypse; Xerces Society.

Living Jewelry

Insects are a frequent motif in jewelry, but sometimes the insect itself is the accessory. The Mayan culture featured living beetle jewelry, though the origins are obscure. Some trace the practice to a Yucatan legend involving a princess and her forbidden affair that led to her lover being sentenced to death. A shaman transformed the man into a beetle, decorated and worn over

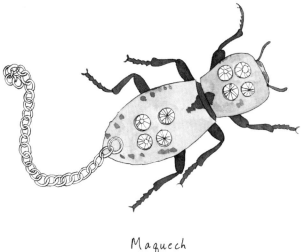

Maquech
Ironclad beetle (with jewels)
~~Zopherus~~ chilensis

the princess's heart as a symbol of their eternal romance. Other scholars suggest this is a tale created by modern Mexican vendors who sell maquech (or makech) curios to tourists. The maquech is a flightless "ironclad beetle," *Zopherus chilensis*.

Colonialism saw the maquech adopted as part of a Victorian-era fetish in which women were expected to wear all manner of organisms to symbolize a reconnection to Mother Nature, lost during the Industrial Revolution. By the 1890s, fireflies blinked from hairdos, and the maquech was decorated with jewels and tethered to the blouse with a golden chain.

The maquech resurfaced in the 1980s; and in 2010, interception of one of the beetles by U.S. Customs drew the ire of the People for the Ethical Treatment of Animals (PETA), who objected to the "bugs in bondage." Indeed, the beetles were being taken from the wild by groups of men ("Los Maquecheros"), who sold them to local artisans who decorated and sold them.

PETA had no such objections to designer Jared Gold's "roach brooch" in 2006, presumably because he rescued the Madagascar Hissing Cockroaches from their normal fate as feeder animals for reptiles. Gold featured his creations on TV's *America's Next Top Model* and sold the bedazzled bugs for $60-$80 each.

Louse Flies

Hunters dressing their kills may be surprised to find small creatures darting through the fur or feathers of the deceased mammal or game bird. They usually remain an enigma, ignored in the preparation of venison or poultry.

The louse flies of the family Hippoboscidae feed on the blood of their host. There are at least 213 species, three-fourths of which are parasites of birds, the remainder occurring on mammals. Some are host-specific, others generalists. Depending on the species, they may be winged, have stubby wings, or no wings at all. Their flattened bodies and sprawling legs, armed with stout, cleft claws, permit them great agility among feathers or fur, and make them difficult to dislodge.

Shockingly, a female hippoboscid hatches one egg at a time, internally. The larva progresses through three instars (an instar is the interval between molts) in what amounts to a uterus. The oviduct of the female fly is

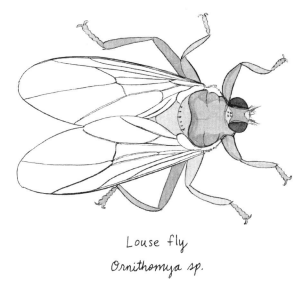

Louse fly
Ornithomya sp.

expanded to accommodate the larva, which feeds on a rich liquid secretion containing two milk proteins and symbiotic bacteria. The mature larva is discharged from its mother ready to pupate. This life stage may be adhered to the host, or hidden in the host's nest, or among leaf litter or other debris. The term for this advanced form of parental care is adenotrophic viviparity.

Mammalian louse flies are known as "keds," the Sheep Ked, *Melophagus ovinus*, being the most familiar and problematic. These small, leathery, bristly, and wingless flies move easily from host to host thanks to the close herding of sheep. Louse flies in general are known to transmit disease-causing bacteria, protozoans, helminths, nematode worms, and possibly viruses. The bird louse *Icosta americana* is a suspected vector of West Nile Virus in North America.

Malaria, Avian

Malaria has long been a scourge of humanity, to the point where we evolved deformed red blood cells (sickle cell disease) to minimize the devastating effects of the *Plasmodium* parasite. We are not, however, the only animals vulnerable to malaria.

Avian malaria is not communicable to humans. It is a different set of more than forty pathogens in three genera: *Plasmodium*, *Haemoproteus*, and *Leucocytozoon*. Over evolutionary eons, most birds have built relative immunity in those regions where it occurs together with its fly vectors: certain mosquitoes, biting midges, and louse flies. It becomes extremely problematic when it appears suddenly among bird populations never before exposed to avian malaria, such as on remote islands, and in zoos and aquaria.

The Hawaiian Islands were a safe haven from avian malaria until the accidental introduction of *Plasmodium relictum* and its vector, the Southern House Mosquito, *Culex quinquefasciatus*, in about 1826. At least ten endemic bird species have since gone extinct. The disease threatens many remaining native forest birds, creeping into habitats at ever higher elevations. One of the effects of avian malaria infection is a reduction in the length of telomeres on the tips of chromosomes. Telomeres are more or less insulators that protect DNA from damage, and their length frequently corresponds to life span of the organism. Birds with avian malaria may not live as long as normal, greatly affecting reproductive capacity.

Birds of the Galapagos and of New Zealand have also suffered from introduced avian malaria. Furthermore, global warming makes more areas susceptible to invasion by malaria vectors normally limited by cooler climates.

Malpighian Tubules

In order to survive, all animals must process food, regulate ion balances, and excrete wastes. Insects have arrived at solutions to these challenges that differ from vertebrates. One of the unique internal organs of most insects is the Malpighian tubules, which figure prominently in filtering wastes and maintaining sodium and potassium ion equilibrium.

To say that the filament-like pouches are the equivalent of the vertebrate kidney would be an oversimplification, but Malpighian tubules are appropriate for the open circulatory system of insects, whereby the entire body cavity is bathed in blood (hemolymph). The number, length, and positioning of the tubules vary considerably from one kind of insect to the next, and they are absent in aphids. Typically, Malpighian tubules are found in a bundle where the midgut ends and the hindgut begins. They produce urine (uric acid), but the hindgut, especially the rectum, reclaims the water and some solutes. That leaves uric acid crystals to be excreted.

Additional functions of Malpighian tubules are also coming to light. It has been discovered that the famous fungus gnat larvae of Waitomo Caves in New Zealand produce their bioluminescent compounds in the Malpighian tubules. Spittle bug nymphs create their "spittle" in those organs. The brochosomes of some leafhoppers are also a product from Malpighian tubules. Malpighian tubules are also the source of silk produced by larvae of lacewings and antlions in constructing the cocoons in which they pupate.

Malpighian tubules are named for Italian scientist Marcello Malpighi (1628–1694), in recognition of his many contributions to microscopic anatomy, histology, physiology, and embryology. He studied both animals and plants, and was the first to recognize that insects breathe through a tracheal system rather than lungs.

See also Bioluminescence; Brochosomes; Spittlebugs.

Mantidflies

Adults of the family Mantispidae look like a bizarre "frankenbug" conglomeration of a mantis and a lacewing, sometimes with waspish overtones. That is the tip of the oddity iceberg.

Mantispids are in the same order as lacewings, and antlions: Neuroptera. There are four subfamilies with 44 genera and 410 species. Adult mantisflies are predatory, and they frequently wait in ambush for another insect to alight on a flower or leaf. Some visit artificial lights at night to feast on other insects.

The life cycle of many species is improbable. Members of the subfamily Mantispinae are, as larvae, predators of spider eggs. The adult female deposits hundreds or thousands of eggs, each on a short, silken stalk, in large batches. The larvae that hatch are "planidia," actively seeking spider egg sacs to penetrate, or they latch singly onto a passing spider. Riding on a male spider is futile, so the larva must transfer to a female during mating of the two arachnids. The larva must enter the egg sac while it is being spun. Once inside, it begins feeding,

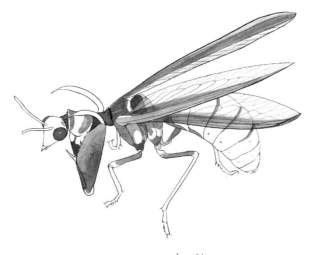

Wasp mantisfly
Climaciella brunnea

molting into its second instar in about a week. That second stage, chubby and lethargic, feeds more earnestly, and molts again in two or three days. The third instar, also grublike, spins a cocoon two to six days later.

Mantispids in the Symphrasinae infiltrate nests of social wasps or bees. The mother mantispid lays her eggs close to a nest of the host. Once inside, her larval offspring prey on the wasp or bee pupae. Getting in is a breeze compared to getting out. Mantispids solve the problem by emerging from their cocoons as pharate adults (basically mobile pupae) that are probably aromatically disguised as members of the host colony. After escaping, they molt again into full-fledged imagoes.

Medicinal Maggots

It may sound like medieval torture, but applying fly larvae to flesh wounds has gone from archaic to an acceptable alternative to antibiotics. Maggot therapy is arguably less painful, less expensive, and a faster route to healing than conventional treatments.

According to some, Genghis Khan kept a ready supply of fly larvae in his traveling war chest, the better to treat his wounded soldiers with rapidity. Napoleon's battle surgeon made use of maggots, too. During the United States Civil War, field hospital personnel noticed that wounds infested with maggots healed faster, and so began piling on more to hasten recovery.

The science behind this phenomenon came later. It turns out that the wound-infesting maggots secrete an antiseptic substance called allantoin. It promotes new cell development and retards microbial infection. The active molecule in allantoin is urea. Yes, the same chemical in urine. Salivary secretions and fecal waste from the larvae apparently both have these sanitation and healing properties. In the saliva is also lucifensin, a novel compound with demonstrable antibacterial qualities.

Faced with ever-increasing incidents of antibiotic-resistant infections, the medical community is turning renewed attention to medicinal maggots as an alternative remedy. MRSA (Methicillin-resistant *Staphylococcus aureus*) has been particularly challenging, and maggots may be the answer.

So effective is maggot debridement therapy (MDT), the technical term for fly-based wound-cleaning, that the U.S. Food and Drug Administration approved maggots as a "medical device" in 2004. There are now companies devoted to the production of maggots expressly for this purpose. The species of choice is often *Lucilia sericata*, a type of blow fly known as a "greenbottle" for the bright, shiny metallic luster of the adult insect. MDT is currently used most often to treat chronic wounds, especially diabetic ulcers.

Merian, Maria Sybilla (1647–1717)

The life and legacy of Maria Sybilla Merian testifies to how far one's passions can take them, and how art can inform science. Even at thirteen, it was apparent that Merian had talent and ambition. She began a journal at that age, her first entry including a stunningly accurate watercolor depicting the metamorphosis of a silkworm moth.

Merian's stepfather was perhaps her greatest artistic influence, but she was no doubt captivated by the many specimens of exotic plants, butterflies, beetles, shells, and other organisms brought back by European explorers and exhibited in cabinets of curiosity. She collected insects herself, and by 1679 she had published her first book on European caterpillars, with fifty plates of illustrations. A second volume followed in 1683.

Her mother passed away in 1690, and a year later Merian was in Amsterdam making connections with the Mayor, Secretary, and at least two scientists, all of whom possessed extensive private natural history exhibits. Perhaps it was those specimens that inspired Merian to undertake a voyage to Surinam (now Guyana) in South America in June of 1699, at age 52, with Dorthea, one of her two daughters. She spent two years rearing caterpillars, many of them brought to her by indigenous residents and African servants. She also observed snakes, lizards, and other wildlife before illness and the stress of the harsh tropical climate forced her return to Amsterdam in September 1701.

Her most famous work, *Metamorphosis Insectorum Surinamensium*, featuring sixty plates, was published in 1705 at her own expense. She had covered those costs as a "ghost artist" for Georg Rumf's *D'Amboinsche Raritei-kamer* (Amboinan Cabinet of Rarities). Her own book went through five editions, the last one in 1771.

Metamorphosis

All insects undergo transformation from egg to adult. Metamorphosis confers distinct advantages that make insects the most successful of all organisms. This is especially true for holometabolous species that pass through an egg, larva, and pupa stage on their way to adulthood.

Each life stage in complete metamorphosis is dedicated exclusively to a specific mission. The larva is the eating and growing stage, while the adult is usually the reproductive and dispersal phase. Immatures do not possess a reproductive system; adults of a few insects lack a digestive system, running on fat reserves accumulated as larvae. Egg and pupa appear externally inert, but they are engines of transformation to the next stage.

Having different life stages also allows a species to partition resources and avoid intraspecific competition. The larval stage feeds on protein-rich food, while the adult stage favors carbohydrates as fuel. Immature stages may even occupy an entirely different habitat from the adult, such as flies that are aquatic as larvae. In temperate climates, different life stages are uniquely suited to diapause and the changing seasons.

Among the few drawbacks of metamorphosis is that much can go wrong during transitions between stages. Even molting from one larval instar to the next can result in deformities. Molting is physically strenuous, and the newly minted animal is soft, pale, exhausted, and more vulnerable to predators. At the level of hormones and genes, everything must proceed flawlessly to properly program the next stage of development.

Our understanding of all the nuances of metamorphosis is an ongoing challenge. Using the few weaknesses in the life cycle against pest insects is a primary objective. Meanwhile, comprehending the requirements for perfect execution of metamorphosis is critical to captive rearing of endangered species.

See also Chrysalis; Cocoon; Diapause; Juvenile Hormone.

Migration

The Monarch butterfly, *Danaus plexippus*, is one of the most heralded of long-distance animal travelers for its annual North American migration (it is non-migratory on other continents), but other insects execute aerial marathons as well.

Few insects that migrate do so with any degree of predictability, or with a common destination. The Painted Lady butterfly, *Vanessa cardui*, is so widespread that its other common name is The Cosmopolite. In the Old World, this species migrates northward in spring from Africa, crossing both the Sahara Desert and Mediterranean Sea to reach Europe. This northbound trip is made by one generation. Once in Europe, successive generations of *V. cardui* continue northward. In autumn,

Painted lady butterfly
Vanessa cardui

their offspring make the return trip. In North America, Painted Lady migrations are infrequent but obvious. In 2017 an epic southbound mob passed through Denver, Colorado, so dense that it showed on radar.

Locusts, the gregarious phase of some normally solitary grasshopper species, migrate by necessity as they exhaust one food resource after another. They take advantage of favorable winds to help propel them over great stretches. Weather also facilitates the movement of many moths, such as the giant Black Witch, *Ascalapha odorata*. A resident of the New World tropics and subtropics, it regularly strays much farther north.

Many dragonflies are built for sustained flight with minimal exertion. The Wandering Glider, *Pantala flavescens*, is in fact the undisputed champion insect migrant. It is known to make the east-west journey from India to eastern Africa, and back, for a total of 11,000 miles or more. Like many insect migrants, they get high enough aloft to be out of sight. The best place to see a stream of Monarchs might be from the roof of your city's tallest building.

See also Kentromorphism.

Miorelli, Nancy

SciComm, short for "science communication," is an increasingly vital aspect of entomology. Advocating for a better public understanding and appreciation of arthropods, and promoting insect conservation, is a high priority with contemporary entomologists like Nancy Miorelli.

Nancy has a master's degree in entomology from the University of Georgia (USA), but her skill at simplifying complex scientific concepts without sacrificing accuracy is perhaps unrivaled. She now practices from her residence in Quito, Ecuador, but manages a global reach through expert use of social media and professionally executed podcasts. She also runs SciBugs Ecuador Tours to the cloud forest, northern coast, Amazon, and Northern Andean Mountains while supporting local guides and communities to foster enduring, cross-cultural relationships. A major earthquake had Nancy soliciting funds from her followers to aid in rebuilding devastated rural villages. Raising consciousness about the interrelatedness of natural ecosystems, indigenous peoples, and local economies is what sets her apart from peers who choose to focus on more specialized niches. Both approaches are equally important and complementary, of course.

Collaborations with friends and colleagues have allowed Nancy to create brands like Ask an Entomologist. She has her own science communication efforts including her tour and education company SciBugs, her Facebook learning community the Sci-Hive, plus her YouTube channel SciBugs. These and other endeavors earned her the Early Career Professional Outreach and Public Engagement Award from the Entomological Society of America in 2017. In her spare time, she crafts jewelry made from the elytra (wing covers) of beetles sourced sustainably in Thailand, and an Ecuadorian palm nut called Tagua. These gems she sells through SciBugs Collections on Etsy. She enjoys painting murals and surfboards, as well as cosplaying.

Mole Crickets

It's a flying crawdad! It's a winged prawn! It's alien spawn, it's . . . a mole cricket? Among the most puzzling insects for the average person are species in the family Gryllotalpidae, so peculiar are they in appearance, and so seldom seen given their subterranean habits.

Eight genera, and more than 80 species of mole crickets are known, collectively found all over the world in tropical and temperate climates. Like true crickets, males use their modified forewings to produce a song, which is amplified and broadcast from a burrow that features a "horn" opening, and a "bulb" deeper underground. The bulb blocks sound from traveling further down the burrow, while the horn acts like a speaker. In some species, the horn is "stereo," with two adjacent funnel-like openings. The front legs are highly modified for digging, the tibia segment typically bearing several large, fingerlike claws (dactyls), and the femur and/or trochanter frequently having a thumb-like projection. The largest species exceed 4 centimeters. Some species are flightless, while in others only females can fly.

Seven exotic mole cricket species have established themselves in the United States (including Hawaii and Puerto Rico), some of which are serious agricultural or turfgrass pests. Meanwhile, the native Prairie Mole Cricket, *Gryllotalpa major*, is a vulnerable species, if not outright threatened or endangered by agriculture, urbanization, and other activities that destroy its habitat.

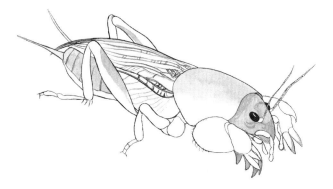

Northern mole cricket
Neocurtilla hexadactyla

Mole crickets are not without their enemies. Various birds, amphibians, shrews, wasps, assassin bugs, carabid beetles, tachinid flies, nematode worms, and fungi and other pathogens attack them. In Thailand and the Philippines, they are considered a delicacy for people. Then again, in Zambia, the African Mole Cricket, *Gryllotalpa africana*, supposedly brings good luck to anyone who sees it.

See also Stridulation.

Moon, Irene

The public face of entomology in the twenty-first century is refreshingly female. One of the more distinguished personas is that of Irene Moon, the stage name of Katja Seltmann. She capitalized on the performance art movement in the late 1990s through about 2010, but many of her "lectographies," songs, and other recordings are archived on the internet.

Dr. Seltmann has not only achieved personal acclaim but also raises the profile for entomology, and for science in general. Her character is a self-described fusion of one of the Lennon Sisters from television's *The Lawrence Welk Show*, and a high school algebra teacher. One critic likened her to a blend of Cruella De Vil (*101 Dalmatians*) and a "strict but nurturing fourth grade teacher." Her bold, authoritative, interactive multimedia performances have taken her on tours in Europe and Australia, as well as around the United States.

Lest one dismiss Seltmann as simply an entertainer, please note that she holds a doctorate from the University of Szeged, Hungary, a master's in entomology from the University of Kentucky, and a bachelor of fine arts from the University of Georgia (USA). She is a published researcher, and currently the Katherine Esau Director of the Cheadle Center for Biodiversity and Ecological Restoration at the Earth Research Institute. She is a strong advocate of STEAM (Science, Technology, Engineering, Art, and Math) curricula, and an admirable role model for girls wishing to pursue a career in those disciplines.

Seltmann's imaginative approach to public education is matched by her dedication to innovation and improvement in the arenas of natural history collections, habitat restoration, and quality of citizen science contributions. Few scientists have worn so many hats with the degree of success and respect that Dr. Seltmann so richly deserves.

Mushi

Among the great tragedies of colonialism, periods of war and regime change, and our monocultural global economy has been the erosion and extinction of indigenous cultural heritage, or the reducing of sacred practices to pop culture or tourist spectacle. Somehow, the practice of keeping *mushi* (insects) has endured in Japan.

It has been a pastime for Japanese children to keep insects, especially "singing" species (*suzumushi*), as pets since the late seventeenth century. Boys and girls would mostly catch their own crickets, katydids, and cicadas, but by 1820 the insects were being bred in captivity for retail sales. Starting during the Meiji period, 1868–1912, vendors erected permanent shops (*mushiya*) to sell all manner of *mushi*, including fireflies and other beetles. By the 1930s, this tradition was waning, and by the end of World War II, *mushiya* were virtually extinct.

Thirty years later, commercial sales of *mushi* experienced a renaissance in department stores, riding a wave of excitement over rhinoceros beetles and stag beetles among young boys. The male beetles were encouraged to do battle in small arenas, or face off in weight-pulling competitions. The market for Coleoptera has continued to expand, and brought with it a return to an appreciation of the seasonal aspect of *mushi*. Experience with live insects was once, and is now again, a means of educating children about the progression of the seasons, mortality, biodiversity, and other concepts of ecological consciousness.

Surprisingly, the metamorphosis of oral *mushi* traditions into products mostly reinforces, rather than detracts from, historical reverence for insects. Electronic facsimiles called *tamagotchi* are enormously popular, as are books and games. Pokémon is a familiar example that has a global following. In Japan, none of these has eclipsed fascination with real insects.

Myrmecophiles

Ant colonies are complex social structures that offer opportunities for other insects to seek food, shelter, and protection from enemies. Ant-loving organisms are called myrmecophiles, and their symbiotic relationships with ants take many forms.

Beetles, flies, true bugs, and wasps are perhaps the most diverse myrmecophiles. Even caterpillars of some butterflies complete their life cycle inside ant nests. Some of these insects are innocuous inquilines, "guests" that impinge little on their hosts, perhaps scavenging in the subterranean refuse dumps of the nest, or soliciting regurgitated food from adult or larval ants. Others prey on ant brood. Mutualistic myrmecophiles typically offer food to the ants in exchange for protection. Aphids, treehoppers, and some butterfly caterpillars provide ants with sweet secretions, while the worker ants serve as bodyguards, chasing off predators and parasitoids.

Those species that exploit ants have evolved elaborate anatomy and behaviors for either escaping detection within the colony, or aborting antagonistic responses. Some simply mimic the odor of their ant hosts. Many beetles that live with ants are heavily armored, with retractable appendages, such that they can withstand bites and stings. Others have modified antennae, or specialized brushes of hairs called trichomes, that dispense chemicals that inhibit attack behavior, stimulate food regurgitation, or otherwise distract the ants or convince them that this oddball creature is a member of the colony.

Nomadic colonies of army ants and driver ants would seem the least likely to host myrmecophiles, but instead they have a rich fauna of freeloaders, groomers, predators, and parasitoids. Rove beetles and phorid flies are the most diverse of these traveling associates. Some of the beetles bear an uncanny resemblance to ants, too.

Nabokov, Vladimir (1899–1977)

The famed novelist, author of *Lolita*, was also an accomplished lepidopterist. Nabokov embodies the idea that one's childhood passion can be accommodated without undue compromise later in life.

By his own account, he became obsessed with butterflies and moths at the age of six. Born into privilege, he had at his disposal quality literature, like Maria Sibylla Merian's *Metamorphosis Insectorum Surinamensium*, accrued by his forefathers. Nabokov was reading entomological journals as a teenager, attempting to reconcile older German standards of taxonomy based on naked-eye morphology with newer, English-driven studies focused on microscopic details. Captivated by mimicry and camouflage, he "discovered in nature the non-utilitarian delights that I sought in art. Both were a form of magic, both were a game of intricate enchantment and deception."

Nabokov arrived in the United States in 1940, moving to Cambridge, Massachusetts where he volunteered at Harvard University's Museum of Comparative Zoology. It was Nabokov who reclassified the rare Karner Blue, a subspecies of the Melissa Blue butterfly, *Lycaeides melissa samuelis*. He was promoted to a paid position as de facto curator of the butterfly collection, and worked in this capacity from 1942 to 1948. Nabokov revised the entire genus *Lycaeides* for North America, and hypothesized, in 1945, that New World tropical blues in the genus *Polyommatus* had a common ancestor in Asia, roughly 10 million years ago. Remarkably, a paper published in 2011 validated this conjecture using molecular DNA analysis and other modern techniques.

While he was alive, Nabokov was hardly embraced by professional scientists, most of whom accorded him amateur status at best, and dismissed his theories and tendency to "split" species. Today, we find more than twenty species of butterflies named after characters in his novels.

Nasute Termites

In the words of the late chemical ecologist Dr. Thomas Eisner, the soldier caste of termites in the subfamily Nasutiterminae (family Termitidae), are "little more than ambulatory spray guns." Each soldier possesses a nasus, a snoutlike projection that shoots defensive chemicals. These "higher termites" probably evolved this weapon specifically for combat against their most lethal predator: ants. In exchange, their mouthparts have atrophied, and they must be fed by workers.

Nasutiterminae are found throughout tropical and subtropical regions of the world. They feed mostly on lichens, dead leaves, branches, twigs, and other plant matter. Some species are subterranean, while others build aerial "carton" nests. In Australia, some species make large mounds. Most build covered foraging trails from their nests to a food source.

Eisner and his colleagues isolated compounds in the poisonous ammunition of soldiers of *Nasutitermes exitiosus* in Panama. They include terpenes that are volatile and aromatic, and highly irritating to other insects. In the mix are also complex diterpenes that give the spray its adhesive qualities and toxicity. This cocktail is ejected as a fine, viscous thread by the soldier termite. It gums up the adversary and usually kills it.

Besides being liquid bullets, the odorous squirts serve two additional functions: They are alarm pheromones that alert worker termites that the soldiers are engaged with an enemy. They also act as recruitment pheromones that rally more soldiers to the site of conflict. What begins as a lone gunman quickly escalates to a 3D firing squad surrounding the aggressor.

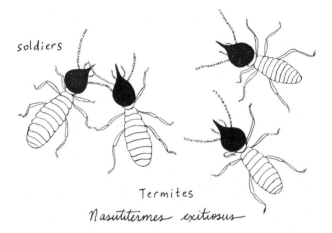

soldiers

Termites

Nasutitermes exitiosus

Even insectivorous mammals like anteaters are quickly routed after breaking into a nest or foraging trail of nasute termites. Some kinds of ants are still specialist predators on them, having evolved strategic "battle plans" to overwhelm and overcome their potent prey.

National Moth Week

Moths suffer many stereotypes: They eat clothing and infest grains. They are dull and drab. Their caterpillars are crop and garden pests. Only a handful of moths fit those categories, but we persist in spraying pesticides and deploying bug-zappers. Enter National Moth Week, an annual, now global citizen science event that takes place the last full week of July.

The concept is the brainchild of David Moskowitz and Liti Haramaty, who have been holding public "moth night" events on behalf of the Friends of the East

Brunswick (New Jersey) Environmental Commission since 2005. The programs have become popular, drawing thirty to fifty people, and countless moths, to each gathering.

Moskowitz and Haramaty, after watching the growth in popularity of the "Moth and Moth-watching" Facebook group, raised the bar in 2011, planning a national week devoted to nightshift Lepidoptera. A year later, public and private events were registered for all of the Lower 48 states and Hawaii, plus Mexico, Europe, and India. The goal? Collect data to improve our understanding of the distribution of moth species and assess the population levels of vulnerable species.

Simply turning on your porch light can attract truly spectacular sphinx moths, giant silkmoths, tiger moths, inchworm moths, and owlet moths. Even urban areas can have a large number of species. Hot, humid nights with no moon are best, but any night will have moth activity.

The website Exploring Nighttime Nature includes a locator page for National Moth Week events. It also features how-to instructions for setting up lights, recipes for "sugaring," and resources to identify the moths you attract. Remember to record images with your camera or phone and upload them to the National Moth Week project for that year on iNaturalist.

See also Underwing Moths.

Neoteny

Some insects would be the envy of the cosmetics industry. They reach sexual maturity while retaining the physical appearance of their youth. This "forever young" condition is called neoteny, and it occurs in vertebrates, too, like the axolotl and other salamanders that remain aquatic and newt-like instead of metamorphosing into terrestrial amphibians.

It is almost always the female that expresses neoteny in insects. She saves metabolic energy in *not* manufacturing wings and other adult features during metamorphosis. That energy she invests in producing more offspring.

Familiar neotenic insects include the railroad worms (glowworms) of the beetle family Phengodidae. Fireflies (Lampyridae) also exhibit varying degrees of neoteny, including larviform females like those of phengodids. Because many species with flightless females are specialist predators, on millipedes in the case of many phengodids, and snails for many fireflies, they are vulnerable to extirpation if their food resource or habitat is compromised.

Most female bagworm moths in the family Psychidae never leave the confines of their cocoons after emerging from the pupa as a larviform adult. It is left for the tiny caterpillars, fresh out of their eggs, to disperse. They do so by issuing silk threads from their mouths and ballooning on the wind to suitable host plants.

One exception to female neoteny is the bark beetle genus *Ozopemon*. These beetles engage in sibling mating, with extreme female-biased sex ratios in offspring as a result of this inbreeding. Males are dwarfed by females, with a flattened appearance, and larviform abdomen. At some point down the evolutionary path, some bark beetles became haplodiploid, the females being able to lay unfertilized eggs that produce males, thus freeing them of the need to mate with their brothers before dispersing.

Niña de la Tierra

The large, wingless insects of the family Stenopelmatidae are known by many aliases, including "Jerusalem cricket," but they are neither crickets, nor from Jerusalem. Distant cousins of grasshoppers and katydids, they are New World sisters to the Old World wetapungas.

In Mexico they are *Niña de la Tierra*, or "Child of the Earth," for the startling resemblance of their large, spherical heads to that of a human infant. The Navajo name *wohseh-tsinni* proclaims the insect's resemblance to an "old bald-headed man." Incorrect translation of Navajo by Franciscan missionaries resulted in "Jerusalem cricket," according to some authorities. Scholar Richard L. Doutt, asserts it may be rooted in substitutes

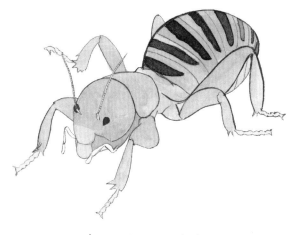

Jerusalem cricket
Stenopelmatus sp.

for profanity in the 1800s. "Jerusalem!" and "crickets!" were popular expletives uttered by children confronted by startling organisms or natural phenomena.

"Potato bug" is another popular epithet for the genus *Stenopelmatus*, especially in California, but Jerusalem Crickets are omnivores and do not single out roots and tubers specifically. Additional names include "skull insect" and "sand cricket."

Stenopelmatus species collectively range from southwest Canada to Costa Rica. Fourteen species were recognized until 2002, when David Weissman at the California Academy of Sciences determined forty-six to sixty-six more species exist, many in isolated sand dune systems in southern California. They can be separated only by behavioral differences. Each species communicates with its own drumming pattern, thumping the abdomen against the ground.

Jerusalem crickets take nearly two years to reach maturity, sometimes extended to five years or more when development is compromised by malnutrition or a parasite. They are prey of owls and bats, any time they surface from their burrows or crawl from beneath stones and other debris on the ground. Digging quickly with their strong, spined legs, or flipping on their backs, gnashing their formidable jaws and kicking their legs, are their go-to defenses.

See also Weta.

Nuptial Gifts
Insects do not have bridal registries, but gifts may be involved in courtship. It can be a way for males to flip the script of female choice that dominates insect mating

systems. Males presenting a gift before, during, or after copulation are communicating superior fitness, securing their genetic contribution to that female's offspring, or at least delaying her eventual receptivity to a competing male.

Adult females of some dance flies in the family Empididae do not hunt, relying on prey captured by males and offered to them. Males gather in aerial swarms, individuals advertising prey items as examples of their prowess. Male balloon flies in the genus *Hilara* wrap their kill inside a silken bubble spun from glands in their front legs. He "gets busy" while she unwraps her meal. Some devious male *Empis* sp. forego hunting and offer an empty wad of their own saliva. Scorpionflies (scavengers) and hangingflies (predators), behave similarly, males furnishing food or saliva.

Males of many katydids supply a spermatophylax at the time of mating. This gelatinous object is delivered with a sperm packet, affording supplemental nutrition for the female's egg output. This can be a heavy investment. The spermatophylax may equal 40% of the male's body weight in male katydids of the genus *Ephippiger*. He is understandably choosy, rejecting a female if she is not sufficiently heavy, indicating low fecundity.

The ultimate gift, of course, is the gift of yourself, and some insects practice sexual cannibalism. Mantids are the most celebrated example. In some species, decapitation of the male by the female results in better . . . performance . . . by removing inhibitory impulses during copulation. In the Sagebrush Cricket, *Cyphoderris strepitans*, the female consumes the vestigial hindwings of her mate after mating, then may gnaw on his forewings, too.

Ootheca

Insect eggs are vulnerable morsels to predators, immobile and protein-rich. Few insects guard their eggs, but most conceal them in some fashion. Mantids and cockroaches package their eggs in a durable mass or capsule called an ootheca.

Technically, other insects manufacture oothecae, the common denominator being that the eggs are held together in a matrix produced from accessory glands in the abdomen of the adult female. Grasshoppers deposit egg pods in the soil. Some tortoise beetles, and at least one stick insect, make oothecae. Some robber flies (Asilidae) and treehoppers (Membracidae), cover their eggs in a frothy substance. Some assassin bugs apply a glutinous substance that holds their eggs together. Opinions vary as to whether these represent true oothecae compared with those of mantids and roaches. Most oothecae are designed primarily to prevent desiccation.

An ootheca is not invulnerable to predators or parasitoids. Ensign wasps in the family Evaniidae are, in the larva stage, predators of cockroach eggs. The mother wasp inserts a single egg into the ootheca, and the larva that hatches consumes the roach eggs within. A female of *Mantibaria*, a genus of platygastrid wasps, alights on a female mantis, and stays aboard until she begins forming an ootheca. The wasp, having removed her wings, is able to penetrate the frothy coating of the mass before it hardens, laying her own eggs in the mantid's eggs.

Several other objects are frequently mistaken for ootheca, and the egg clusters are often assumed to be cocoons, or galls. Certain fungi and slime molds look much like the product of an insect. Oozing foam insulation and related synthetic substances may at first glance look like an insect egg mass, too.

Ovipositor

A revolutionary addition to insect anatomy was the evolution of egg-laying organs that allowed female insects to conceal their vulnerable ova within a substrate, or upon or inside a host. More astounding still was the conversion of that organ to a venomous weapon.

An ovipositor can be as simple as the telescoping segments of the abdomen, as in many grasshoppers, most beetles, many flies, and other insects. More advanced examples are found in dragonflies and damselflies, katydids, crickets, true bugs like cicadas and leafhoppers, some flies, and sawflies. These are complex, multisegmented modifications of the terminal abdominal segments, specialized to penetrate soil, or plant or animal tissues. These blade-, spear-, or whip-like appendages may look like stingers, but true stings are usually retracted inside the abdomen when not in use.

Advanced ovipositors are not single, fixed appendages, but composites of two or more valves operated independently by abdominal muscles at the base of each. Additional valves may form a two-part sheath around the ovipositor, as they do in ichneumon wasps, for example. They can also serve as bracing structures, akin to an oil well derrick, when "drilling" into dense wood to reach a host larva.

The sting and its associated venom gland likely evolved to deliver toxins to temporarily paralyze a host, making it easier to lay an egg on or inside a compliant victim. From there, the sting and venom became a way to permanently paralyze a host, allowing the wasp to transport it and conceal it from animals that would otherwise pilfer the prize. Eventually, with the advent of sociality, some wasps, ants, and bees turned the sting into a weapon for self-defense and, more importantly, defense of the helpless brood: eggs, larvae, and pupae inside the nest.

Periodical Cicadas

Magicicada is the genus of seven uniquely American periodical cicadas. Three species are on a 17-year cycle, and four are on a 13-year cycle. The predictability of their synchronous emergences, and sheer numbers of individuals, make these insects truly spectacular.

Scattered populations emerging the same year constitute a "brood." There are twenty-three broods, each represented by a Roman numeral. Some broods are widespread, others highly localized. One is extinct. Each brood usually includes more than one species. Like most cicadas, the males produce a loud song with tymbal organs located in air-filled chambers in the abdomen. Cicadas are the only insects that generate sound with a true percussion instrument. The collective din of a *Magicicada* emergence has been compared to the sound of a flying saucer in an old sci-fi movie.

Aside from the riot, and the smell of decaying bodies when they expire, periodical cicadas are remarkably benign. Females can damage twigs and branches in the course of drilling into a tree to lay eggs. The result is "flagging" as foliage beyond the egg scars turns brown and dies. The nymphs that hatch from the eggs fall to the ground and tunnel into the soil. There they will

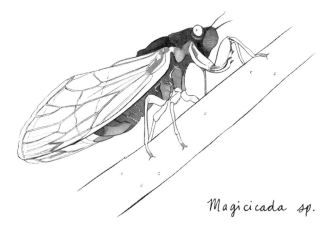

Magicicada sp.

spend the next thirteen- or seventeen years sucking sap from plant roots.

Why the lengthy, odd-numbered cycle? This is still a mystery. Perhaps it was an evolutionary solution to a devastating parasite. Even so, there is a fungus that takes a heavy toll on the adults. Birds, mammals, reptiles (even copperheads), and other animals feast on them. So do people. The Iroquois harvested them as food. Sauteed, the nymphs apparently taste like shrimp, or canned asparagus. Check out the "Cicada Mania" website for more on *Magicicada*.

Pesticide Treadmill

Humanity has clung stubbornly to the idea that chemical insecticides are the best way to treat troublesome insect species in agricultural crops. This is in spite of documented problems, not the least of which is resistance to toxins in the targeted pest.

Dr. Robert van den Bosch coined the term "pesticide treadmill" in his 1978 book *The Pesticide Conspiracy*. The principle is that application of pesticides reveals resistance to those toxins in some members of the target pest population. Those individuals survive to produce a generation with even more resistant individuals. This requires heavier applications of the insecticide, or a newer, more potent treatment. It becomes a vicious circle in which humanity never gains the upper hand.

Besides killing the intended targets, insecticides kill predators and parasitoids that would normally achieve at least supplementary control of the pest. In the absence of enemies after a pesticide application, the resistant individuals of the pest species thrive until natural enemies rebound, if they ever do. This was proven in a 2019 study published in the journal *Oecologia*, concerning mosquitoes in Costa Rica that breed in bromeliad plants that catch and retain rainwater. Treating the reservoirs with dimethoate killed some, but not all, mosquito larvae. Those that survived fed better in the absence of competing siblings. The damselfly larvae that would have preyed upon them were totally eradicated by the insecticide.

Even "successful" pest control campaigns can have unforeseen consequences. Suppression of one pest may liberate another. During the late 1940s, cotton plantations were treated with DDT, carbaryl, and azinphosmethyl to control Boll Weevil in south Texas and neighboring states to the east. The result was the emergence of Cotton Bollworm and Tobacco Budworm as the new leading cotton pests in the early 1950s.

See also Biocontrol; Integrated Pest Management.

Pheromones

A common assumption about pheromones is that they are purely sexual "fragrances" emitted by females to attract males. In reality, they are diverse, serving a remarkable array of functions between individuals of the same species.

The word is from the Greek *pherein* (to carry or transfer) and *hormon* (exciting), so translates roughly to "excitement carrier." There are two categories: Releaser pheromones are the familiar type that are neutral to the emitting individual, but elicit a behavioral response in the receiver. Primer pheromones cause a physiological change in the receiver and mostly affect physical development, especially among social insects.

Pheromones are complicated molecules that contain a staggering amount of information. To a female moth, the manufacture of a pheromone is also less costly than the energy expended in searching for a male. A male, for his part, may need to convince her he is a worthy mate. He may emit his own pheromone, communicating his superior fitness, or ability to deliver a defensive chemical during copulation, which she will use to protect her eggs.

Many insects use aggregation pheromones to attract more individuals of one or both sexes to a specific location, as some bark beetles (Curculionidae: Scolytinae) do. Pheromones rally large numbers of pine beetles to overwhelm the defenses of a tree that normally drowns boring beetles in resin flows. Only when resins are depleted can the beetles successfully colonize. At that point, the beetles emit anti-aggregation pheromones to repel additional beetles.

Social insects use a variety of pheromones. Trail-marking pheromones laid down by scout ants returning from a successful foraging expedition stimulate following behavior in other workers, reinforcing the trail to the food. When a nest of bees is under assault, the act of stinging releases an alarm pheromone that alerts the entire colony to the threat and recruits more workers to the battle front.

See also Hair-Pencils.

Phoretic Copulation

Some male insects literally sweep their mates off their feet. Species in which the female is wingless, but the male is winged, frequently engage in the insect equivalent of the "mile high club."

This unique form of mating has arisen several times in solitary wasps, and requires males to be considerably larger than females. Members of the families Thynnidae (flower wasps), Mutillidae (velvet ants), and Bethylidae practice phoretic copulation, and some wasps in the Tiphiidae likely do so as well.

Flightless female thynnid wasps from Australia benefit considerably in this scenario, because the male furnishes his mate with food in one of three ways: directly from trophallaxis, whereby he regurgitates food into her mouth; by storing food under his "chin," from which she feeds; or by transporting her to a source of nectar. She solicits his advances by crawling up a stem or stalk and adopting a "calling" posture, releasing a pheromone that attracts him.

Physical engagement of the couple in thynnids and mutillids begins with the male grasping the female's thorax, just behind her head, with his jaws. In some velvet ants, he may stroke her thorax with his front legs. Contacting the female's abdomen with base of his middle and hind legs is a trigger for her to orient her abdomen for coupling of their genitalia. Females are immobile passengers during flight, and may be oriented to face forward, or rearward. Mating itself may take place during flight, or once the couple lands.

Tandem flights of other insects do not qualify as phoretic copulation. Both sexes participate in locomotion, or either sex takes the lead. Both phoretic copulation and tandem flights can be a form of mate-guarding. Males are protecting their genetic investment by denying the female an opportunity to mate with another male.

Pine Processionary Caterpillars

Larvae of the Pine Processionary, *Thaumetopoea pityocampa*, are known for defoliating forests in southern Europe, northern Africa, and Central Asia. Their irritating hairs cause agony for anyone who brushes against them.

Fabre's accounts of this species in *The Life of the Caterpillar* (1916) are highly captivating, though we have learned more since his observations and crude experiments. In summer, the female moth deposits roughly 250 eggs in a cylindrical cluster, camouflaging them with scales from the tip of her abdomen. The caterpillars that hatch progress through five instars before the final molt to a pupa stage.

The marching behavior, single-file, head-to-tail, begins in earnest in the third instar (an instar is the interval between molts). By then, in late autumn, they have built communal tents of silk, home to hundreds of individuals.

Pine processionary larvae
Thaumetopoea pityocampa

At night, they follow each other, laying down a strand of silk from their mouth, and a trail-following pheromone from the rear. Eventually, each one settles to feed on its own patch of needles. In late spring, the caterpillars descend the tree, wandering over the ground by day in search of pupation sites. This conspicuous parade differs from their foraging behavior. Instead of following a pheromone or silken path, they are stimulated by the setae on the rear of the caterpillar in front of them. They pupate singly in loose cocoons, a few inches under the soil surface.

Do not let curiosity get the better of you should you come across a column of caterpillars. Each one is covered in setae (hairs) that are barbed, easily penetrating the skin. Besides mechanical irritation, the hairs contain at least seven allergens that cause painful contact dermatitis. In hypersensitive persons, they can trigger anaphylactic shock, a life-threatening episode.

See also Fabre, Jean Henri; Pheromones; Urticating Hairs.

Pollinators

"Save the bees" has become a popular cause driven by the apiculture (beekeeping) industry. Conservationists point out that native, solitary bee species, plus other pollinators, are in worse trouble.

Most insects are "flower visitors," there for nectar refills. Pollination may or may not be accomplished in the endeavor to refuel. A few insects, including some short-tongued bees, are nectar *robbers* that slit the throats of blossoms that have deep, narrow corollas, to reach the nectar while bypassing anthers, pistils, and stamens. The best pollinators are those that eat pollen, or actively harvest it, which include bees, beetles, a few wasps, some moths, and *Heliconius* butterflies.

Besides the colors that attract us, flowers may sport "nectar guides," markings visible in the ultraviolet spectrum that only insects can see. We may love the fragrances of flowers, but a few produce a stench that mimics decaying flesh, to lure flies as pollinators. Many flowers set elaborate traps to entice pollinating insects, or have kinetic mechanisms that deliver pollen to a precise location on the insect.

Many agricultural crops rely on insect pollinators. If you love figs, thank a wasp. Like chocolate? Tip your hat to a biting midge. The list goes on. The scale of modern industrial agriculture often outstrips the capacity of native pollinators to provide, however, endangering both cultivated and wild ecosystems. Habitat destruction for other human wants such as timber, fossil fuels, and urbanization adds to the perils for pollinators, along with climate change, introduced species like honey bees, and continued use of pesticides. Individual consumers and landowners can help by practicing responsible shopping, and by landscaping and rejuvenating local ecosystems with native plants.

See also Fig Wasps; Hummingbird Moths; Pseudo-copulation; Xerces Society; Yucca Moths.

Pseudocopulation

The coevolution of plants and insects has led to some unusual, if not downright embarrassing, relationships. Some flowers dupe male wasps into attempting to mate with them, which only ensures fertilization of the floral kind. This is pseudocopulation.

The most celebrated example of wasps accidentally fornicating with flowers surrounds the mirror orchid or looking-glass orchid, *Ophrys speculum*. The plant occurs in the Mediterranean region, along with its sole pollinator, a scoliid wasp, *Dasyscolia ciliata*. To the male wasp, the flower looks, smells, and even feels like a female wasp. The blue color mimics her iridescent wings. The fragrance evokes the pheromone the female wasp releases to attract suitors. Alighting on the flower, the male interprets tactile stimuli as further evidence he has

Scoliid wasp
Dasyscolia ciliata

settled upon a receptive mate. Attempted copulation triggers the flower to adhere one or two pollen packets (pollinia) to the male wasp's face. Fooled again, the wasp deposits the pollinia in the next flower he visits.

Many other orchids employ this deceptive strategy, from Japan to South America. In Australia, two species of tongue orchids in the genus *Cryptostylis* are pollinated exclusively by males of the ichneumon wasp *Lissopimpla excelsa*. So convincing is the flower that the male may even ejaculate onto the flower. Another Australian orchid, *Caleana major*, masquerades as a female pergid sawfly in the genus *Lophyrotoma*.

Milkweeds also package pollen in pollinia, but they can be pollinated by any strong insect able to extract the tacky units without getting fatally stuck to them. As is the case with the orchids, it is mostly large wasps that are the most efficient pollinators of milkweeds.

See also Pollinators.

Queen

Social insects are not as obsessed with royalty as we are, but colonies do have a division of labor whereby one or more females are dedicated solely to reproduction. We call them queens. In the case of social wasps, bees, and ants, a queen is a "gyne."

Some insects show similarities to a queen-based lifestyle but do not have true queens. A "stem mother" aphid may hatch in spring from an egg laid in winter. Her parents reproduced sexually, but she will create the next generation through parthenogenesis (the production of viable offspring without mating). She thus starts a new colony of aphids, but there is no division of labor.

Paper wasp colonies can be established by one or more female foundresses, but a single female eventually asserts dominance through physical bullying. Daughters of gynes are initially workers with undeveloped ovaries that cooperate in rearing their siblings, and eventually the next generation of gynes.

Ants, bees, and wasps have a unique mode of sex determination called haplodiploidy. This means that female offspring are produced from fertilized eggs, but males are the product of unfertilized eggs. Worker females, normally nonreproductive, can lay eggs that produce males. This may happen if the queen dies.

Some ant species establish colonies in a fashion similar to paper wasps, but many species have colonies that grow to enormous populations with several nests and multiple queens. Dispersal of ants, and the founding of new colonies, is accomplished by periodically liberating swarms of males and unmated gynes from existing colonies. Both sexes are winged ("alates"). Unrelated colonies swarm synchronously to avoid inbreeding.

Termites are the only social insects with a male complement to the queen. A "king" mates repeatedly with his queen throughout their lifetimes, which can be more than forty years for queens of some tropical species.

Rain Beetles

On the west coast of North America, there live beetles with a life cycle akin to periodical cicadas. The rain beetles in the genus *Pleocoma* are twenty-six species found collectively from southwest Washington to northern Baja California. Most are isolated in mountains or valleys.

Rain beetle
Pleocoma puncticollis

The family Pleocomidae was once a subset of the scarab beetles, and they resemble "June bugs" in appearance, shiny reddish brown or black, with a shaggy coat of hairs on the belly. The adults do not feed, having vestigial mouthparts and no digestive tract. Males fly in search of females, burning fat reserves accumulated as larvae. They may have as little as two hours of flight time. The female is larger, and flightless. She remains inside or near her larval burrow and emits a pheromone to draw suitors. In most species, all adult activity takes place in late fall, winter, or early spring, often in pre-dawn hours, or at dusk, especially in the aftermath of heavy rains or snowmelt.

After the brief orgy of mating, males die and females descend into the depths of their tunnels. It may take months for her eggs to mature, but she eventually deposits forty or fifty eggs in a spiral pattern within her burrow. The grubs hatch in about two months and may bore to a depth of 3 meters in their search for the roots of trees, shrubs, and grasses upon which they feed. Like cicada nymphs, they grow leisurely, molting seven or more times. Rain beetles take eight to fifteen years to reach adulthood.

Our human relationship to them is ambivalent. They can be pests in orchards where the grubs munch on roots. Conversely, they are so unique that "Rufus" rain beetle was once nominated for state insect of Oregon.

See also Periodical Cicadas.

RIFA

Historically, human colonists have invaded foreign soils and made pests of themselves. Insects have, too. RIFA stands for Red Imported Fire Ant, *Solenopsis invicta*, one of the worst invasive insects in the southern United States. The distinction is important: four *native* fire ant species occur in the U.S.

Native to South America, RIFA arrived in Mobile, Alabama in soil used for ship ballast, in the 1930s or early 1940s. The ants are so well adapted to seasonal flooding in their tropical homelands that there, colonies withstand deluges by forming living rafts of their own bodies, floating to new nest sites. RIFA thrives in disturbed habitats, from vacant lots to parks, gardens, and farms. Stepping in a nest results in hordes of ants scrambling up your ankles to bite and sting. A red welt with a central white pustule is their signature.

RIFA destroys seeds and damages crops. They feed on honeydew, so they enhance populations of aphids and scale insects. Their mounds damage farm equipment. They scavenge dead insects and vertebrates, but they also prey on live insects, helpless bird chicks, and baby rodents.

Beginning in 1957, when Congress approved an eradication program, we enthusiastically jumped on the pesticide treadmill, trying one chlorinated hydrocarbon (chlordane and its ilk) after another. Next came Mirex, banned by the Environmental Protection Agency in 1976. All we achieved was elimination of competing, native ant species, making life easier for the invader.

The Achilles heel of the ant may turn out to be one of its natural enemies, such as a fungus, nematode worm, or, most satisfying of all, a fly, *Pseudacteon obtusitus*, that decapitates ants in the course of its development. Meanwhile, RIFA has invaded Australia, New Zealand, and several countries in Asia, plus several Caribbean islands.

Riley, Charles Valentine (1843–1895)

If any entomologist deserves superhero status, it would be C. V. Riley. Besides advancing his profession, he helped rescue French vineyards and American citrus crops from disaster.

Born in London, Riley attended a French boarding school, then immigrated to the United States at age 17. At 25 he was appointed State Entomologist of Missouri. His influence extended far and wide. He lobbied to secure state money to fight an infestation of LD Moth in Massachusetts. Then, from 1868 through the mid-1880s, Riley collaborated with French scientists to thwart the Grape *Phylloxera*.

Riley suspected the aphid-like *Phylloxera* had been exported from America to Europe. Riley and his "Americanists" eventually convinced their French "purist" counterparts that *Phylloxera* was indeed the source of withering vines, not an "effect" of another malady. The ultimate remedy was grafting French vines onto American rootstock. Riley was heralded as a hero, receiving the Legion of Honor, the highest award of the French government.

Riley's ego led his detractors to refer to him as "the General." Still, he was selected as Chief of the U.S. Entomological Commission in 1876, a committee formed to address an outbreak of Rocky Mountain Locust. In 1878, Riley was appointed as the first entomologist of the United States Department of Agriculture.

In 1887, the introduced Cottony Cushion Scale was decimating the fledgling citrus industry in California. Riley sent a colleague to Australia to search for natural controls. He selected the Vedalia lady beetle. In late 1888 and January 1889, live beetles were received in Los Angeles, cultured inside tents erected over trees. By the summer of 1889, citrus growers declared a resounding victory.

Today, Riley is memorialized by a foundation that bears his name, and remembered not only as an entomologist but also as a naturalist, artist, writer, and Renaissance man.

See also Gypsy Moth; Rocky Mountain Locust.

Rocky Mountain Locust

The story of *Melanoplus spretus* is a cautionary tale. Once the most abundant insect in North America, it is now extinct.

The density of locust swarms, which peaked in the late 1800s, is unfathomable. The largest, in 1875, was estimated at 3.5 trillion insects, covering 198,000 square miles. It was part of an outbreak that lasted from 1874 to 1877, costing $200 million in damages ($116 billion today). From Nevada to Missouri, and Texas to Canada, Rocky Mountain Locust was a biological terrorist.

Outbreaks occurred every six or seven years but endured one to three years, leaving widespread poverty in their wake. State and local jurisdictions offered bounties for harvests of grasshopper eggs and nymphs. Many contraptions were invented to kill them, including horse-drawn flame-throwers. Farmers were forced to diversify from vulnerable wheat to alfalfa, peas, and beans that were less appetizing to locusts.

By the end of the 1800s, swarms became smaller and sporadic. On July 19, 1902, the last living specimens of *Melanoplus spretus* were documented. Some believed the conversion of fields to alfalfa starved the insects. Others were convinced that the extirpation of bison from the Great Plains compromised locust habitat. Similarly, it was postulated that driving indigenous peoples from the plains removed historical management tools like fire.

The mystery was solved when scientists revisited the geographic distribution of the grasshopper. Intermountain valleys of the Rockies were the only "permanent zone" for the grasshopper. The gold and silver rush in the mid-1800s had transformed those areas into agricultural landscapes. The soil containing grasshopper egg pods was plowed and irrigated. It was this devastation of its breeding ground that did the species in.

Knife Point Glacier in Wyoming is perhaps the largest burial ground for deceased specimens, but like most glaciers it is fast retreating.

See also Kentromorphism.

Schmidt Sting Pain Index

Entomologist Justin O. Schmidt has been labeled the "King of Sting" for converting his unpleasant encounters with insects into quantifiable rankings. It is the kind of logical solution to a problem that characterizes Schmidt's intense curiosity and innovative approach to scientific inquiry.

In graduate school at the University of Georgia (USA), Schmidt found himself studying *Pogonomyrmex* harvester ants as models of stinging insects. Evaluating the relative toxicity of various species was straightforward, but a way to assess pain levels was sadly lacking. In a 1983 paper, published with two other authors in *Archives of Insect Biochemistry and Physiology*, Schmidt included a pain scale of 1–4, with the Western Honey Bee, *Apis mellifera*, as the standard "2". Enough people can relate to the sting of a honey bee that they can extrapolate to pain of greater or lesser intensity. Schmidt's colorful descriptions of stings, and imaginative comparisons to other horrific pain-inducing scenarios, has won him many devotees.

Far from being an eccentric, or mad scientist, Schmidt is a scholar of the highest order. He generally avoids inducing stings intentionally, gaining knowledge of the perception of various insect stings through random episodes. Not content with the sensations alone, Schmidt seeks to comprehend the biochemistry that produces the results. He has found that separate venom compounds are usually responsible for pain, and actual damage, respectively. Pain is an indicator of impending or concurrent damage, Schmidt stresses in his explanation of how stings function. Social wasps, ants, and bees are the most toxic, creating a threshold at which a vertebrate predator must abandon its efforts to destroy the colony for brood (soft larvae, pupae) or food stores (honey), to live to hunt another day.

See also Venom.

Screwworm

If you have never heard of the Screwworm, *Cochliomyia hominivorax*, it is due to the success of one of the most unique methods of insect control ever conceived.

The maggot stage resembles a screw, hence the name. Female flies deposit several hundred eggs each near wounds on mammals. The tiny larvae that hatch burrow into the exposed flesh. Infestations can cause the death of livestock and wildlife due to toxic compounds released by the maggots, or from secondary infection. Native to the tropical and subtropical Americas, screwworm was a horrific pest of cattle and sheep across most of the southern United States in the 1950s.

At the U.S. Department of Agriculture, Edward F. Knipling and Raymond C. Bushland had been plotting a devious, nuclear-age pest control method ever since publishing their theory in 1937. SIT, for "sterile insect technique," involved rearing, sterilizing, and releasing vast quantities of a given pest to crash population levels. The screwworm was a test subject, on Sanibel Island, Florida, and Curacao. It was a resounding success.

In 1962, Congress allocated money for eradication of screwworm in Texas. A plant was built in Mission that produced up to 140 million flies per week. Pupae were bombarded with gamma rays from radioactive Cobalt-60 to achieve sterilization. The pupae were then packaged in special boxes and loaded aboard modified cargo planes that dispersed the pupae over the landscape. The sterile male flies mated with wild females, yielding no offspring.

The factory in Mission was shuttered in 1984, after helping push the "barrier zone" deep into Mexico. Today, the edge of the fly's range is now the isthmus of Panama. SIT continues to be applied to screwworm, and other pests, particularly certain species of fruit flies (family Tephritidae), and tsetse flies in Africa.

See also Integrated Pest Management; Tsetse Flies.

Seed Dispersal

Human beings plant farms, orchards, and gardens, but what is responsible for seeding forests, prairies, meadows, and other wild landscapes? Birds and mammals, obviously, but also insects, particularly ants.

Entomochory is the scientific term for seed or spore dispersal by insects. Myrmecochory applies to ants only. More than 4,000 species of angiosperms, representing more than seventy families of plants, rely on myrmecochory for seed dispersal. To entice their transport, the plants attract ants with special fleshy structures or seed coatings called elaiosomes. The ant typically carts the entire package to the nest. There it distributes the elaiosome to nest mates. These bodies are loaded with lipids (fats) the ants have difficulty securing from other food sources.

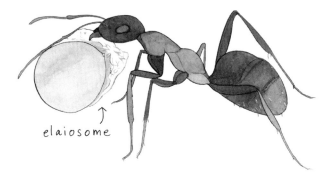

elaiosome

Once stripped of the yummy parts, the seeds are usually deposited by the ants into their "garbage" chamber, the equivalent of a compost heap. There, the seeds enjoy an optimal microclimate and nutrient source. Other ant species return the seeds to the surface of the soil. In habitats with regular fire events, ants can position seeds at precisely the right depth to avoid incineration, but still receive the heat required for germination.

Other insects disperse seeds, too. Scarab beetles bury countless seeds in the animal dung they covet. In Australia, stingless bees disperse seeds in the course of harvesting resin from Cadaghi (*Corymbia torelliana*). The bees discard the seeds, but not until after they have flown some distance.

The ability of ants to disperse seeds can be compromised by clearing of land for timber and farming. Second-growth forests in North America can have higher concentrations of non-native earthworms, which deprive the forest floor of leaf litter that provides cover for ants. Invasive slugs also eat elaiosomes but don't disperse the seeds.

See also Ecosystem Services.

Sericulture

No substance of animal origin has had such a profound impact on civilization as silk. Cultivation of *Bombyx mori* has resulted in the only domesticated insect that no longer exists in the wild.

Folklore describes empress His-ling Shih fumbling a cocoon into her hot tea in 2640 BC and learning she could unwind the thread. Archaeological evidence suggests silk was woven as far back as 6,000 to 7,000 years ago. Traditional Chinese methods of farming the caterpillars forbid loud noises, strong fragrances, and untidiness, sure to spoil the synchronicity essential for successful rearing.

The caterpillars feed exclusively on mulberry leaves, increasing their weight by 10,000–14,000 times in five instars, four molts, and twenty-five days. Cocoons are exposed to hot air or steam to kill the pupae. The cocoons are then dipped in hot water to dissolve sericin, the gummy substance coating the silk protein fibroin. Only then can the single thread, 1,200–1,600 yards long, be unwound onto a spool.

China managed a monopoly on silk through many dynasties, but legend suggests Byzantine emperor Justinian I, in AD 522, dispatched a group of Persian monks to smuggle silkworm eggs. They did so inside hollow canes, returning to Constantinople to initiate a western silk empire.

Begun in the early second century BC, the Silk Road was a network of overland paths and associated ocean passages, from China to India and the Mediterranean. At its peak during the Tang Dynasty, AD 618–907, jade, spices, and gold were traded along with ideologies of Buddhism and Islam.

Today, the major players of sericulture are constantly changing in rank. Historically, China and Japan were leading producers of silk, but France, India, Italy, Russian Federation, Pakistan, Uzbekistan, and Brazil have joined the fray, among others. China remains the overwhelming supplier.

See also Cocoon.

Snow Insects

Logic would suggest that no self-respecting, cold-blooded insect would be caught out in the dead of winter, but while most insects are in diapause, others take advantage of the lack of competition and predators.

"Snow fleas" are actually springtails, *Hypogastrura nivicola* and *H. harveyi*, which can form dense mats on snow, especially as it begins to melt, and around the base of tree trunks throughout the year. They feed on rich nutrients provided by fallen leaves and other decomposing organic matter. They withstand the cold by means of a protein with high concentrations of the amino acid glycine, a self-produced antifreeze.

Snow flies are among the more obscure of snow insects. They are small, lanky, wingless crane flies in the genus *Chionea*. They survive mostly in the void between the snow blanket and the warmer soil beneath, but make forays onto the snow surface to find mates. They are most comfortable between 0° and −4° Celsius and peak in activity between October and November, and again in February–March. The adults may live as long as two months, without feeding, at least not on solid foods. The sixteen known North American species have few winter predators.

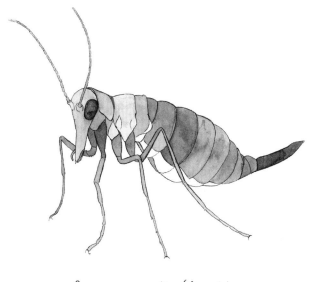

Snow scorpionfly (female)
Boreus sp.

Most bizarre of all snow insects are snow scorpion-flies in the family Boreidae. At only 2–5 millimeters, they are hardly conspicuous. Boreids chew on leafy parts of mosses and liverworts as adults and larvae. Adult females are wingless. In males the wings are hardened, reduced to straplike stubs, each with a sharp terminal spine. He uses them to grasp and maneuver the female, the fairer sex mounting *him* instead of the other way around. The family name, and most common genus, *Boreus*, refers to the boreal forests where they are most commonly found.

See also Hexapod; Grylloblatids.

Spittlebugs

A familiar summer pastime for children is blowing soap bubbles. Some young insects blow bubbles, too, but not as a leisure activity. The frothy spittle commonly seen on garden and field plants protects the immature insect from enemies and the drying effects of the hot sun.

The spittle is produced by nymphs of insects known as froghoppers in the superfamily Cercopoidea, composed of five families with more than 360 genera, and about 2,600 species. They are closely allied to cicadas, treehoppers, and leafhoppers. Most spittlebugs feed as both nymphs and adults on succulent, herbaceous vegetation such as grasses, clover, and alfalfa, but other

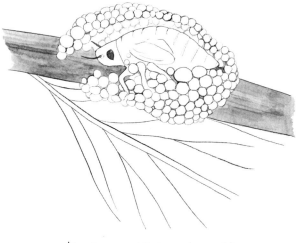

Meadow spittlebug (nymph)
Philaenus spumarius

species occur on shrubs and trees, including conifers. They pierce the plant stems with a segmented rostrum, extracting sap.

It is an excess of these imbibed plant juices that makes up much of the "cuckoo spit." The surplus fluids bypass the midgut by means of a "filter chamber" and are expelled through the anus, together with a kind of liquid wax manufactured by the Malpighian tubules. This waxy substance thickens the lather and probably helps prevent the bubbles from bursting. A groove on the underside of the insect channels air from the spiracles (breathing holes) toward the posterior of the spittlebug. A valve near the anus regulates the air flow. Vigorous movements of the abdomen draw in air, then churn the bubbles into a froth. The groove and valve also allow the bug to breathe by protruding its posterior just outside the spittle mass.

As adult insects, froghoppers abandon their foam homes for a more active lifestyle. Although they can fly, their preference is to jump away if danger threatens.

See also Malpighian Tubules.

Spotted Lanternfly

Made in China, now at large. That describes the Spotted Lanternfly, *Lycorma delicatula*. The polka-dotted planthoppers suck the sap of trees and shrubs, especially Tree-of-Heaven. Their indiscriminate appetite makes them a threat to forests, fruit orchard crops, and grape vineyards. At least 103 species of plants are agreeable hosts. Spotted Lanternfly (SLF) appears to be native to China, India, Bangladesh, and Vietnam, but it has been introduced into South Korea and Japan as well as the United States.

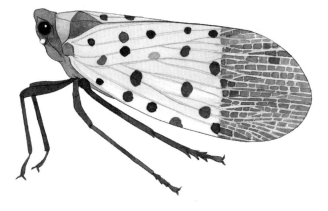

Spotted lanternfly
Lycorma delicatula

SLF was first detected in the United States in Berks County, Pennsylvania on September 22, 2014. It is spreading through movement of materials on which egg masses are attached. The flat, gray, linear clutches can be adhered to any smooth surface and are easily overlooked. Vehicles, outdoor furniture, firewood, flowerpots, and countless other objects can have egg masses, so inspection is key.

The life cycle is one of incomplete metamorphosis. Egg masses are the overwintering stage. Nymphs hatch in spring and are initially black with white spots, becoming red in the fourth instar before reaching adulthood. Adults appear in midsummer, with mating and egg-laying occurring in autumn.

While feeding they excrete copious amounts of honey-dew that coat leaves of the host plant and any understory foliage, breeding sooty mold. SLF is most easily observed at dusk and after dark, gathering in large groups on trunks and stems of plants.

Spotted Lanternfly is poised to become a global agricultural pest. In North America, it is capable of spreading throughout the northeast and mid-Atlantic states, westward through the Great Lakes to eastern Kansas, at minimum. The interior valleys of central California are also at risk, as well as orchards in eastern Washington and Oregon. Much of Europe, southern Australia, the cape of South Africa, and southern South America are also vulnerable.

Stridulation

Sound production by insects takes many forms, but the most common is the rubbing of one body part against another, an action known as stridulation.

The loud "calling songs" of male crickets and katydids are generated by the highly modified front wings. Katydids have a hardened ridge of teeth ("file") near the base of the left wing, and a sharp "scraper" on the opposing wing. The two are rubbed rapidly against one another to produce the song. Most crickets are "right-handed," the file on the right wing, scraper on the left. The male also produces a rivalry song to aggressively assert his territory to trespassing males. When a female approaches, he switches to a softer courtship song. Both sexes hear through openings in the front legs that each contain a sensitive membrane called a tympanum.

Grasshoppers, by contrast, usually stridulate by rubbing a row of pegs on the inside of the hind femur ("thigh") against raised veins on the front wing. The tympanum in grasshoppers is located in a cavity on each side of the front of the abdomen.

Bark beetles (Scolytinae), some burying beetles (Silphidae), and a few velvet ants (Mutillidae) also stridulate as part of courtship. Mutillids normally use stridulation as a defense, the auditory equivalent of aposematism. They do so by rubbing one abdominal segment against another. Other insects that make warning sounds include assassin bugs (Reduviidae) that rasp the tip of the rostrum against transverse ridges on the underside of the thorax. Many longhorned beetles (Cerambycidae) also rock their heads, scraping their "necks" against ridges on the inside of the top of their thorax. Such unexpected noises can startle a would-be predator into dropping the insect.

Swarm

Swarm, the noun, is usually a benign gathering in the insect world. Swarm, the verb, can be a terrifying event from the perspective of humans faced with a mob of insects. We usually perceive large numbers of insects as threatening, when in fact most swarms are simply a nuisance.

Honey bees offer a perfect example of both. A swarm results from division of a colony, whereby roughly half the worker force splits from the established colony to found a new one, together with a newly emerged queen. The swarm forms a temporary cluster while scout bees fly in search of a new nest site. A bee swarm can also be a "congregation area" of drones, an aerial ballet of male bees for purposes of mating with new queens. Neither situation is the least bit menacing. Honey bees will, however, swarm any predator intent on pillaging their honey stores and/or helpless brood. Swarm, the entity, is peaceful. Swarm, the action, is necessarily fierce.

Not every localized abundance of insects constitutes a swarm. Most are aggregations related to a food resource, nesting area, diapause, or synchronized emergence. Lady beetles form dense masses during diapause, which may take place in winter or, in arid regions, during the summer as the insects migrate vertically from hot valleys into cooler mountains.

There are never "swarms" of beloved species like butterflies. No, we say there is a swarm of wasps, locusts, or flies. That swarm of "mosquitoes," by the way, is more likely a swarm of midges, harmless flies distantly related to mosquitoes. Lake Malawi in Africa hosts spectacular swarms of lake flies, *Chaoborus edulis*. The swarms are so dense and enormous that they can be mistaken for smoke plumes or interpreted as living tornadoes.

Tequila Worm

Men are prone to do silly things in the name of bravado, like eating the worm at the bottom of the tequila bottle. The thing is, it is not a worm, nor tequila.

The liquor is mezcal. Tequila *is* a mezcal, but it must be a minimum of 51% blue agave to qualify as tequila, compared with mezcals produced in other proportions from a variety of agave plants. One theory of the origin of the "worm" was as a means of distinguishing other mezcals from tequila.

Cossid moth (larva)
Comadia redtenbacheri

Two kinds of insect larvae are used. One is the larva of the Agave Weevil, *Scyphophorus acupunctatus*. The feeding and egg-laying activity of the adults can open the plant to infection by fungi and other stem- and leaf-rotting agents that make life better for the larvae, which mine the roots and stems.

The worm of choice is *Comadia redtenbacheri*, caterpillar of a cossid moth. In Spanish, they are known as *gusanos rojos*, feeding in the heart of the agave, the part roasted and distilled to make mezcal. Young caterpillars are gregarious and migrate toward the rhizome, where they disperse and mature. Their development can take more than a year but varies by individual. The last instar larvae are reddish and release a volatile, stinky compound as protection against enemies. This suggests why the presence of a worm in the bottle is said to change the taste of the liquor.

Tequila was granted protected status on October 13, 1977, and labeling mezcal with a preserved larva was no longer necessary. Some traditions die hard, or are resurrected for marketing, so a few brands with larvae persist today. An intact specimen signals a purer beverage, but ingesting the creature will do nothing for your sexual prowess, nor initiate hallucinations. Sorry.

See also Entomophagy.

Tok-Tokkies

Deathwatch beetles smack their heads to make sound, but "tok-tokkies" of southern Africa are the butt-thumpers of the beetle world. Both sexes pound the underside of the abdomen against the ground, with surprising rapidity, to communicate.

The common name of these peculiar darkling beetles is onomatopoeic, meaning a word derived from another that is created by mimicking the sound being interpreted, like "bang," "pow," or "knock." In this instance it is "tok-tok," and the beetle making the sound a tok-tokkie.

All species of tok-tokkies are in the family Tenebrionidae, but there is no formal classification beyond that, as several, but not all, genera from the subfamily Pimeliinae, tribe Sepidiini, use the a**-slamming habit for communication. *Psammodes*, *Phrynocolus*, and *Ocnodes* are the most renowned for this. They occur in arid habitats, are flightless, and have a dense exoskeleton. Unlike many darkling beetles, they do not have chemical defenses, relying on their armor, and their surprising running speed, to avoid and repel predators. Their ability to withstand extreme heat during daytime hours also exposes them to fewer predators than there are in the cool of the night. Water conservation is an even greater priority, and the edges of the elytra (wing covers) fit snugly into a continuous groove along the periphery of the abdominal sternites (ventral segments) to reduce water loss in respiration.

Tok-tokkies are scavengers, feeding on a variety of dry, organic, windblown detritus, including dead invertebrates, plant matter, and feces.

Several other beetles are known for tapping behavior, including the tenebrionid genera *Eusattus* and *Coniontis*, which inhabit deserts and prairies in North America. Several genera in the jewel beetle family Buprestidae drum their abdomens against the surface of logs, branches, or trunks during courtship.

See also Deathwatch Beetle.

Tsetse Flies

The "king of beasts" in Africa may not be the lion, but biting flies in the genus *Glossina*. Historically, the thirty-one species and subspecies of tsetse halted the southerly advance of Islam, and impeded colonialism.

The flies are notorious for spreading microorganisms that cause sleeping sickness, but it is nagana, a related illness afflicting livestock, that is economically devastating. The collective ranges of *Glossina* form sub-Saharan "fly belts" that fluctuate between wet and dry seasons, preventing cattle herds from reaching optimal grazing

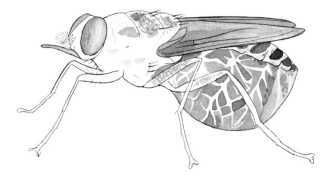

Tsetse fly after a blood meal

Glossina sp.

lands. The tsetse thus stands between a wildlife paradise and its conversion to a colossal cattle ranch, since wild mammals are immune to nagana but act as reservoirs for the trypanosomes that cause the disease.

Glossina is divided into savanna species, forest species, and riverine species. Not all bite humans, but they will do so through clothing. Gambian Sleeping Sickness, accounting for more than 98% of reported cases, is typically chronic, death occurring years after the date of infection. Rhodesian Sleeping Sickness is more virulent, and death occurs in weeks or months. Thirty-six African nations are affected, but only 997 new cases were recorded in 2018, according to the World Health Organization.

The tsetse life cycle is similar to louse flies. Females hatch one egg at a time, nursing a single larva in a "uterus," fed from "milk glands." The larva passes through three instars in about nine days. It is then "born," and promptly digs into the soil where it pupates. An adult emerges in four or five weeks. One female can produce five to eight offspring over her lifetime of three or four months.

According to the fossil record, North America had tsetse flies about 35 million years ago. It is suspected that equines evolved stripes due to tsetse and other biting flies. The bold, vertical patterns disorient the flies at close range, causing them to miss their target or collide awkwardly.

See also Florissant Fossil Beds; Louse Flies.

Underwing Moths

Underwing Moths
Hide, flash, and fly is the survival strategy of owlet moths in the genus *Catocala*, known as underwing moths. The insects combine camouflage, aposematism, and powerful flight to escape daytime predators.

The genus name is Greek and translates to *kato* for "below" and *kalos* for "beautiful," a perfect description for a moth with forewings mottled in shades of gray or brown, and hind wings strikingly banded in black and orange, pink, red, yellow, or white, or solid black. Despite their large size (3.5–8.5 cm wing span), the moths are perfectly camouflaged against the bark of tree trunks, or beneath rock overhangs and other sheltered situations during the day. Disturb one, and it will flare the front wings to reveal the bright hind wings, then take flight. This "startle display" distracts and disorients most would-be predators.

Catocala are favorites with collectors, but are not drawn to lights at night as frequently as most moths. They do have a weakness for sweet baits, so the collecting

Underwing moth
Catocala sp.

method most often used for them is "sugaring." This involves painting the trunks of trees with a goopy concoction, then visiting the trapline after dark. Each person has their own secret recipe, but the basic ingredients are molasses, brown sugar, an overripe fruit, and an alcoholic beverage. Brown sugar, stale beer, and a mashed banana is the poor man's formula, but others swear by various combinations of liquors and fruit. Allow the mixture to age a couple of days before applying.

There are around 270 species of *Catocala* at last count, found mostly in temperate climates of North America and Eurasia. Most underwings fall into two categories: those that feed as caterpillars on oaks, or those that feed on poplar and willow.

See also National Moth Week.

Urticating Hairs

While most insect larvae are harmless, many caterpillars are covered in setae that provoke unpleasant reactions in humans. Hypersensitive immune responses can even lead to anaphylaxis, a life-threatening event.

True setae in the obnoxious offenders are fine, barbed hairs that detach easily from the insect's body, like porcupine quills. They can lodge in skin, but are usually airborne, so inhalation of the setae is the greatest hazard. Notable examples of true setae include Pine Processionary caterpillars, tussock moths, and the Browntail Moth, some giant silkmoths, and prominent moths. Prior to pupating, the caterpillars weave their setae into their cocoons for protection.

There is no secretion from these setae, so the proteins inherent in the cuticle of the hair must act as irritants or allergens. Those inflammatory properties persist long after the hairs depart the caterpillar; and reactions typically occur twelve hours or more after exposure. Prolonged, repeated exposures can be lethal. Browntail Moth has killed at least two laboratory researchers studying the species long term.

Modified setae are not as numerous as true setae, and are typically arranged in clusters. They are thicker, hollow, and with basal glands that secrete a toxin wicked into the bristle when it breaks, exuding onto or into the attacking predator. The larvae possessing such setae are classified as venomous by some authorities. They are common in several families of moths.

Venomous spines are stout setae, often branched, and with venom glands within them. Such stinging caterpillars are nothing to trifle with. Classic examples include giant silkmoths like the Io Moth, and the Saddleback Caterpillar, Monkey Slug, and their kin. The caterpillars of flannel moths (aka puss moths, asps, family Megalopygidae), have venomous spines concealed under long, thick setae. Diabolical.

See also Pine Processionary Caterpillars; Venom.

Venom

Ouch! A surprising array of insects deliver venom through stings or bites, but the definition of venom is expanding to include almost any substance that causes damage or physiological change in the recipient.

Venom, and delivery systems for it, have evolved at least twenty-nine times in the class Insecta. The most familiar are wasps, bees, and ants, but there are stinging caterpillars, too, and, stunningly, a stinging beetle. The adult longhorned beetle *Onychocerus albitarsis* has a venom gland in the sharpened, hollow terminal segment of each antenna. Like caterpillars, the sting is used for self-defense against vertebrate predators. Predatory true bugs, flies, and lacewings and their kin possess paralytic enzymes at the very least. Blood-feeding insects inject anticoagulants and other toxins.

In wasps, venom probably evolved to temporarily paralyze a host, making it easier to deposit eggs on or inside it. Wasps traditionally considered nonvenomous have been found to inject chemicals that facilitate fungal growth in plant hosts, or which act in concert with specific viruses to take control of the host animal's nervous system. By this measure, all wasps may be venomous. Social wasps, bees, and ants may or may not use

Saddleback caterpillar
Acharia stimulea

their stings to kill prey, but all use them in defense of the helpless eggs, larvae, and pupae (brood) inside the nest. The composition of venom thus varies according to its intended purpose.

Stinging caterpillars can be as potent as wasps and ants, if not more so. Most notorious is *Lonomia obliqua*, a gregarious giant silkworm found in South America. In Brazil and Venezuela, it has been responsible for hundreds of deaths. Its venom, delivered through nasty spines, contains anticoagulants that can cause a hemorrhagic syndrome.

See also Schmidt Sting Pain Index; Urticating Hairs.

Vespa mandarinia (aka "Murder Hornet")

The Asian Giant Hornet is, according to the media, the latest and greatest biological threat to North America. One reporter dubbed *Vespa mandarinia* the "murder hornet" for the propensity of this species to raid honey bee hives, annihilating worker bees in the process.

The panic began in September 2019 when a nest of *V. mandarinia* was discovered in Nanaimo, British Columbia, and destroyed. Subsequently, a single deceased hornet was recovered in Blaine, Washington, USA, in December. Genetic analysis determined that the dead specimen had a separate origin from the Nanaimo colony. Additional live specimens were observed or collected in spring and summer of 2020 in Langley, British Columbia, and Custer, Bellingham, and Birch Bay, Washington.

On October 22, 2020, a nest was discovered in Blaine in a hollow tree, the occupants suctioned out by entomologists donning protective suits and wielding special vacuums. An estimated 500 individuals occupied the nest, including roughly 200 new queens that were destined to disperse for winter diapause in scattered locations. In early November, individual hornets were captured in Abbotsford and Aldergrove, BC. Clearly, the Asian Giant Hornet is of ongoing concern in the Pacific Northwest, but only there.

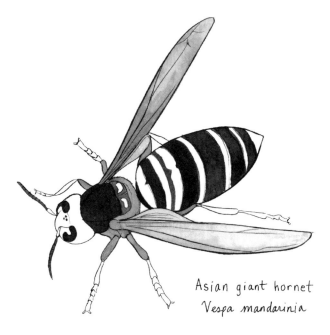

Asian giant hornet
Vespa mandarinia

By inciting widespread fear, media sensationalism has triggered wasteful expenditures of money and manpower in states where the hornet is likely to never appear. Responsible journalism would highlight lax inspections of shipping containers at international ports. Honest, brave writers would suggest that a deregulated global economy threatens to undermine some industries (apiculture in this case), and destroy native ecosystems, through accidental and intentional importation of exotic species. Instead, various news outlets compete for clicks, re-tweets, and market share with inflammatory headlines.

Viviparity

Most insects achieve reproductive success by laying large numbers of eggs to offset the inevitable death of most of their daughters and sons. A few protect their genetic investment in some fashion, and one strategy is birthing "live" offspring. Retaining small numbers of offspring within the female's body obviously reduces opportunities for predators and parasitoids to kill them.

The scientific term for live birth is viviparity. It runs the gamut from eggs hatching internally and the nymphs or larvae expelled immediately (ovoviviparity), to carrying the larval stage "to term," short of the pupal stage (adenotrophic viviparity). A surprising variety of insects have adopted these means of reproducing. Aphids are a familiar example, birthing nymphs that grow quickly and repeat the process through parthenogenesis. In some species of aphids, one generation will reproduce this way, while another will reproduce sexually and lay eggs. Many flightless cockroach species produce live nymphs instead of the usual oothecae. Additional species that are viviparous include some scale insects, earwigs, a few barklice, true bugs in the family Polyctenidae, and some thrips. Even a select few beetles in the families Carabidae, Staphylinidae, Tenebrionidae, Micromalthidae, Cerambycidae (*Borneostyrax cristatus*), and Chrysomelidae exhibit viviparity.

Viviparity can help a species stay one step ahead of the competition, too. Flesh flies (Sarcophagidae), for example, get a head start in exploiting the ephemeral resource of an animal carcass by larvipositing, laying tiny maggots instead of eggs. Satellite flies, a subset of flesh flies, also larviposit, reducing the risk of facing their stinging wasp and bee hosts inside the confines of a nest. The tiny maggots are less conspicuous, and able to quickly wriggle inside the nest from its entrance.

See also Louse Flies; Ootheca; Tsetse Flies.

Water Striders, Marine

The saltwater realm is generally considered the domain of crustaceans, the marine complement to the terrestrial and freshwater insects. Considering that crayfish, crabs, woodlice, and other crustaceans have invaded rivers, streams, lakes, and ponds, as well as many land habitats, it should come as no surprise that insects have invaded the strongholds of crustaceans.

The marine environment includes ocean beaches, intertidal zones, mangrove swamps, and saltmarshes. A great diversity of insects can be found in these places, but truly pelagic insects are nearly nonexistent. The exception are certain water striders in the true bug family Gerridae that literally surf the open ocean. Five species in the genus *Halobates* are known to inhabit tropical and subtropical seas. They were first discovered during a transoceanic expedition of the Russian ship *Rurik* between 1815 and 1818. The insects have sprawling legs that distribute their weight over a greater area to prevent them from breaking the surface tension of water. They feed on other tiny animal life that falls onto the water surface.

How do *Halobates* keep from drowning during storms? Two layers of fine, dense hairs, one long and one short, help make the insects water-repellent. They also groom frequently, applying a waxy secretion from thoracic glands to their bodies to render them essentially waterproof.

Sea sKater
Halobates sp.

Being small (4–5 millimeters) and lightweight (about five milligrams) also helps them to maintain buoyancy. *Halobates* are flightless but athletic, capable of vaulting themselves off the surface of the water, even somersaulting, to avoid both breaking waves and predators attempting to grab them from below or above.

Weta

The Giant Weta, *Deinacrida heteracantha*, lays claim to being the world's heaviest insect, the record being 70 grams (~2.5 ounces). This is one of eleven species of wingless, cricket-like animals in the genus *Deinacrida* found exclusively in New Zealand and neighboring islands. Despite its bulk, the Giant Weta lives in the forest canopy.

Weta is short for "wetapunga," the Maori name for the Giant Weta. It was the Maori who unfortunately introduced the kiore, or Polynesian rat, to the islands occupied by the wetapunga. The lumbering insects, formerly without predators, quickly fell victim to the rodents. By the 1700s, European colonists had arrived, contributing exotic rodents, pigs, goats, deer, and opossum. The weta's instinctive defense, to forcefully kick with its heavily-spined hind legs, is apparently no deterrent to the introduced predators.

Adult male wetapungas range in size from 5.2 to 5.7 centimeters, females 6–7.3 centimeters (nearly 3 inches). They are large enough that some researchers now track them with radio transmitters. It takes a long time to get that big, and the entire life cycle exceeds two years. Eggs take roughly four months to develop before the female oviposits. The nymph progresses through ten instars in about 18 months, and adults live an average of four months (females) to seven months (males).

Today, the Giant Weta is listed as "vulnerable" by the International Union for Conservation of Nature (IUCN), and nationally endangered in New Zealand. The original population is confined to Little Barrier Island, now devoid of all mammals since rats were eradicated in 2004. Captive breeding was initiated in 2008, with some success. Progeny of the captured individuals were released on Mouora and Tiritri Matangi islands in 2011 and 2014, respectively. In late 2020, the species was reintroduced to Northland, after a 180-year absence.

See also *Niña de la Tierra*.

Wigglesworth, Vincent Brian (1899–1994)

As father of the study of insect physiology, Sir V. B. Wigglesworth deciphered the holy grail of entomological mysteries: how hormones regulate insect growth, metamorphosis, and reproduction.

Wigglesworth's childhood was spent in outdoor exploration and collecting butterflies and moths in England. At the age of eight he purchased a microscope to examine the scales on the bodies of his specimens. From age seven he was attending boarding schools; he served briefly at the end of World War I.

"VBW" enrolled at the University of Cambridge in 1919, earning a B.A. in Anatomy, Physiology, and Zoology. This landed him a scholarship and eventually a studentship from the Department of Scientific and Industrial Research. From 1922 to 1924, his focus was on vertebrate metabolism. He received his M.D. in 1929, and the Raymond Horton Smith Prize for best thesis.

In 1926, VBW was invited to be a lecturer at the London School of Hygiene and Tropical Medicine at London University. There, he lucked into a perfect research subject in the kissing bug, *Rhodnius prolixus*, a species reared at the school. In short order, Wigglesworth made great strides in the understanding of insect development. He penned *Insect Physiology* in 1934, followed by *Principles of Insect Physiology* in 1939. The latter went through seven editions.

In 1945 Wigglesworth returned to Cambridge University. In 1952 he was elected to the Quick Chair of Biology. His diligence yielded more than three hundred publications in his career. Wigglesworth was honored with appointment as a CBE (Commander of the Most

Excellent Order of the British Empire) in 1951, and the Queen knighted him in 1964. He officially retired from Cambridge University in 1967. He capped his resumé with receipt of the Gold Medal for Insect Morphology at the 1992 International Congress of Entomology.

See also Juvenile Hormone.

Wolbachia bacteria

Insects have many symbiotic relationships with microbes, but the most common, diverse, and *perverse* are with *Wolbachia*. An estimated 76% of all insect species are affected by various strains of the bacteria. One individual insect can be infected by more than one strain.

The most common effects of *Wolbachia* infection are in reproduction, where it can determine the sex of the host offspring, and their survivability. When a male and female insect mate, each infected with a different strain of *Wolbachia*, the result is no viable offspring. This is called cytoplasmic incompatibility. The male is a dead-end host for *Wolbachia*. Consequently, when only one parent insect is infected, progeny are heavily skewed toward females. This reaches an extreme in many parasitoid wasps, where *Wolbachia* facilitates parthenogenesis. Ordinarily, female wasps are the product of sexual reproduction, whereas males come from unfertilized eggs and have only one set of chromosomes compared to a female's two sets. The bacterial infection permits a female to produce fertile female offspring without mating.

Another bizarre outcome from *Wolbachia* infection can be feminization of genetic male offspring, such that they develop into functional females. This is known in two crambid moths, *Ostrinia furnacalis* and *O. scapulalis*, and a leafhopper, *Zyginidia pullula*. It may also occur in other insects, but the mechanism is not fully understood, and may vary from one species to another. Other scenarios include the outright killing of male embryos by *Wolbachia*, or increased fecundity of females infected with *Wolbachia*.

Wolbachia also have the potential to make a host insect inhospitable to other microbes. This is of great interest to entomologists fighting insect vectors of infectious diseases. The bacterium could be genetically engineered to interrupt the cycle of the disease-causing organism within its vector host.

Xerces Society

It may sound absurd to suggest that there are species of "bugs" in need of protection, but there are many. Thanks to forward-thinking individuals, there are now organizations devoted to invertebrate conservation.

The first of these to be founded was the Xerces Society, named after the extinct Xerces Blue butterfly, at the time thought to be the only North American insect ever exterminated by man. Lepidopterist and author Dr. Robert Michael Pyle founded the Xerces Society in 1971 as a loose organization of dedicated entomologists and naturalists. In 1973 it launched the journal *Atala*, and in 1974 the newsletter *Wings*. Pyle's wife, Sarah Anne Hughes, spearheaded the organization of Fourth of July Butterfly Counts, modeled after the Christmas Bird Counts of the Audubon Society. The first butterfly counts were held in 1975 at 29 locations.

Xerces blue
Glaucopsyche xerces

The Monarch Project was the first major endeavor of Xerces, aimed at protecting the butterfly's wintering sites in California and Mexico. Melody Mackey Allen joined Xerces as a volunteer fundraiser for the effort, then became its first paid employee in 1983. She was named Executive Director in 1985.

Since then, the Xerces Society has diversified considerably, becoming a leader in the conservation of all pollinators, aquatic invertebrates, arthropods endemic to old-growth forests in the Pacific Northwest, and more. The society empowers rural farmers and urban homeowners to change the management of croplands and gardens to promote biodiversity. Xerces impacts are felt from the United States to Costa Rica, Madagascar, and beyond.

The butterfly counts are now run by the North American Butterfly Association, as they have been since 1993.

The professional journal *Atala* ceased publication in 1987, but Xerces has put out several books and countless educational publications since, both in hardcopy and online.

See also Endangered Insects.

Yucca Moths

The famous symbiotic relationship between moth and yucca is less than half the story. The family Prodoxidae is full of surprises.

In the classic scenario, a female of the genus *Tegeticula* harvests pollen from the anthers of the flower stamens, using tentacle-like protrusions on her palps to ball up a wad of pollen. She then flies to flowers of another yucca and crawls to the base of the pistil to pierce the ovaries of the flower and lay her eggs within. Afterward, she ascends the pistil and kneads the pollen into the stigma. Her eggs hatch into caterpillars that feed on some of the fully-developed seeds. They eventually exit the plant to pupate in the soil, emerging as adult moths the following spring or summer.

Females of some *Tegeticula* species lack pollen-packing paraphernalia. These "cheater" moths are active later in the season and lay eggs in the developing fruit around the seeds. "Bogus yucca moths," *Prodoxus* sp., behave similarly to the cheating *Tegeticula* species. The caterpillars of some *Prodoxus* species feed on the fruit, others on the flower stalk. In the course of eating, secretions from the caterpillars cause the plant to develop a hardened capsule around the insect. Inside, the pre-pupa caterpillar can remain in diapause for years, even decades, if winter weather conditions are not optimal.

Then there are other pollinating species in the genus *Parategeticula*. In the life cycle of *P. pollenifera*, the female moth scrapes pits into the petals or stalk of the flower and lays her eggs. Her emerging caterpillars each tunnel into immature seeds, which initiates a cyst of abnormal tissue to grow in place of a few seeds. It is the cyst on which the larva feeds.

See also Pollinators.

Zombie Lady Beetles

A tiny female wasp sneaks up on an unsuspecting lady beetle and drives a single egg into its body using her spear-like ovipositor. The wasp larva that hatches starts feeding as an internal parasite. The story is just beginning.

With her egg, the wasp, a braconid named *Dinocampus coccinellae*, also deposits a virus. This virus is called *D. coccinellae* Paralysis Virus (abbreviated DcPV). The larva that hatches also carries the virus.

Once the wasp larva is ready to enter the pupa stage, the virus migrates to the beetle host's brain where it asserts a semi-paralyzing effect. Alternatively, it may be the beetle's immune response to the virus that causes the brain damage. The wasp larva exits the beetle and spins a silken cocoon beneath it. While the host is helpless to move of its own free will, the virus triggers periodic twitches in the beetle, making it a zombified bodyguard of the wasp cocoon.

This symbiotic wasp-virus tag team may be a widespread phenomenon. Polydnaviruses (PDVs) are a category of viruses integrated into the genomes of many wasps in the Braconidae and Ichneumonidae. Replica-tion of the virus takes place in the female wasp's reproductive organs, but its virulence is unleashed on the host. DcPV is by contrast a heritable iflavirus that replicates in the *host's* neural tissues. PDVs can exert similar mind-control effects on hosts, though.

Dinocampus coccinellae is found worldwide, and it is not specific as to which lady beetles it uses as hosts. Female beetles may be preferentially targeted, and the feeding activities of the wasp larva may even sterilize the host, meaning that even if a victim recovers, which about a third of the beetles do, it will be unable to reproduce.

Zoos, Insect

Living zoological collections traditionally exhibit charismatic megafauna such as tigers, elephants, and giant pandas. Only fairly recently have these facilities begun to fulfill their obligation to represent the full spectrum of biodiversity by displaying terrestrial arthropods.

The crude beginnings of insect zoos date to small collections resulting from colonial exploration, such as those at the Jardin de Plantes, Muséum d'Histoire Naturelle et Ménagerie in Paris, France in 1797. The first permanent fixture was probably the Insect House at London Zoo in 1881. The Artis Zoo in Amsterdam, Netherlands debuted its exhibit in 1898, and several zoos in Germany followed suit through the early 1900s. Two world wars interrupted the advancement of invertebrate displays, save for seasonal exhibits at the Bronx Zoo in New York, Goddard State Park in Rhode Island, and Brookfield Zoo in Chicago in the 1930s and 40s.

The Sherbourne Butterfly House in Dorset, England set a new trend in 1960, with emphasis on the most charismatic invertebrates: butterflies. Creepy cockroaches, beetles, and other insects are easier to breed in many instances, but harder to sell to zoo accountants and the visiting public. Most butterfly houses rely on regular importation of butterfly pupae from tropical butterfly farms. This model employs indigenous people and otherwise poverty-stricken populations in those countries. Other insects cultured behind the scenes require only one importation of wild stock, or purchase or trade from another zoo.

The logistics of insectariums, whether part of an existing zoo, aquarium, or a stand-alone enterprise, are complex. The facility must meet strict guidelines for containment of exotic species such that they cannot escape and become invasive. They must continue to justify their missions as public attractions, fulfill public education mandates and, increasingly, participate in invertebrate conservation programs.

Afterword

Curiosity. That is the overriding character attribute of entomologists. An entire set of encyclopedias would still fail to do justice to the subject, but the reader has hopefully come away with a deeper appreciation for insects, and a desire to learn more.

Assuming mankind does not continue to decimate entire ecosystems around the globe, insects will forever be a subject of intrigue and astonishment. As the digital age expands, the impact of "amateurs" on entomology will become ever greater, and more respected.

It is in the acceptance and promotion of *human* diversity that entomology perhaps faces its greatest challenges. We must acknowledge that much of our historical knowledge of insects has come through colonialism. Likewise, the traditionally patriarchal culture of all sciences no longer serves us. To their credit, the current generation of entomologists recognizes this and is making strides to correct it.

Meanwhile, transition from a pest-killing mentality to an interest in the conservation of insect diversity must overcome the inertia of profit-making enterprise, and accommodate endeavors that are necessary for the survival of entire ecosystems. The private homeowner, plant nursery manager, landscape company, city planner, rural farmer, and others, can all have a hand in promoting a new paradigm of coexistence with our six-legged friends. It can happen. It has to.

Acknowledgments

The author first thanks Robert Kirk of Princeton University Press for inviting him to author this volume, in the fine tradition of *Fungipedia* by Lawrence Millman. Special thanks to artist Amy Jean Porter for her delightful and accurate illustrations, and for sharing her imaginative interpretations with the author to inspire creativity. Lucinda Treadwell's expert copyediting made the manuscript much more consistent, comprehensible, and livelier still.

The author's mentors are too numerous to mention by name, and the list grows longer with every project. Suffice that the community of entomologists is always eager to aid science communicators. Finally, my spouse, Heidi Eaton, remains my rock, the one with the steady job, who not only tolerates my irregular writing career, but also encourages it.

Useful References

Berenbaum, May. *Bugs in the System*. Addison-Wesley. 1995.

Eisner, Thomas. *For Love of Insects*. Harvard University Press. 2003.

Evans, Howard E. *Life on a Little-known Planet*. E.P. Dutton. 1968.

Heinrich, Bernd. *In a Patch of Fireweed*. Harvard University Press. 1984.

Marlos, Daniel. *The Curious World of Bugs*. Penguin Group. 2010.

Paulson, Gregory S., and Eric R. Eaton. *Insects Did it First*. Xlibris. 2018.

Stewart, Amy. *Wicked Bugs*. Workman Publishing. 2011.

Tallamy, Douglas W. *Bringing Nature Home*. Timber Press. 2007.

Teale, Edwin Way. *Grassroot Jungles*. Dodd, Mead & Company. 1969.

Waldbauer, Gilbert. *Insights From Insects*. Prometheus Books. 2005.